The World at your feet

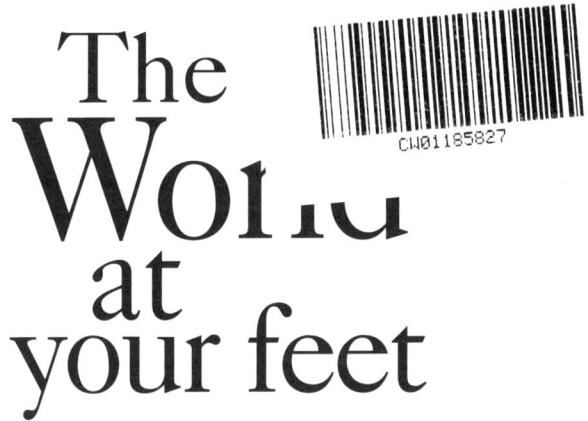

A traveller's guide for women
in mid-life and beyond

Nina Hathway

BOOKS

© 1996 Nina Hathway
Published by Age Concern England
1268 London Road
London SW16 4ER

Editor Gillian Clarke
Design and production Eugenie Dodd Typographics
Copy preparation Vinnette Marshall
Printed in Great Britain by Bell & Bain Ltd, Glasgow

A catalogue record for this book is available from the British Library

ISBN 0-86242-189-6

All rights reserved; no part of this work may be reproduced in any form, by mimeograph or any other means, without permission in writing from the publisher.

Every effort has been taken to ensure the accuracy of the information contained in *The World at your feet*, but Age Concern England and the author cannot accept any responsibility for any errors that may exist, or for changes that occur after publication.

CONTENTS

FOREWORD BY *JILL DANDO* 6
ABOUT THE AUTHOR 7
ACKNOWLEDGEMENTS 8

1.
Decision time 9
First things first 10

2.
Travelling alone, with a companion or in a group? 12
Going alone 12
Two's company 20
The group 23
Having second thoughts about travelling? 28

3.
Travellers with special needs 29
Pre-trip tips 31
Medical requirements 33
Insurance 34
Booking your trip 34
Main travel options 36
Individual disabilities 39

4.
Before you go 41
    Getting a good deal 42
    Finding accommodation 43
    Your holiday budget 43
    Essential documents for foreign travel 46
    Insurance 47
    Taking a car abroad 50
    Looking after your home 51
    Packing 53

5.
Staying safe 57
    Sensible precautions 58
    Tips for travelling alone 61
    If you are robbed 63

6.
Staying well 65
    Before you leave 65
    Prepare early 67
    Immunisations 68
    Common travellers' complaints 71
    Essential medical supplies 75
    Safe eating and drinking 77
    Hygiene 77
    If you're ill 77

7.
Holiday possibilities **80**
  The retreat 80
  Low-cost breaks 81
  Rent a property 82
  Pamper yourself 83
  Study/special-interest holidays 83
  Living with a family 85
  Camping and caravanning 85
  A home away from home 86
  Life on the ocean waves 87

8.
The world at a glance **89**
  Europe 91
  Asia 106
  Africa 111
  Australasia 115
  North America 116
  South America 119

USEFUL ADDRESSES 121

FURTHER READING 129

ABOUT AGE CONCERN 131

PUBLICATIONS FROM AGE CONCERN BOOKS 132

INDEX 135

FOREWORD

The world has no walls. How I've noticed that over the years I've been travelling! These days, thanks to lower prices and better means of transport, there are very few places untouched by the inquisitive traveller. Consequently, older people and women in particular have fewer qualms about getting on a plane and broadening their perspectives.

Since all too few travel books cater for women and almost none for the older age groups, I'm delighted that *The World at your feet* provides a practical guide to overcoming problems and enjoying the experience – whether you're planning an independent trip, making the most of a group holiday or swapping homes with people from another country.

Happy reading and happy travels!

Jill Dando
Presenter of BBC 1's Holiday

ABOUT THE AUTHOR

Nina Hathway is an experienced journalist who writes about women's issues and travel for a wide range of newspapers and magazines.

After living in places ranging from Arabia to Germany when she was a child, Nina has since lived and worked abroad in countries such as the USA and Norway. She continues to travel whenever possible, although these days she is less likely to take off with only a backpack and summer work permit.

ACKNOWLEDGEMENTS

This book is dedicated to Eva and John Hathway.

My thanks go to David Moncrieff who commissioned me to write this book, Richard Holloway and Evelyn McEwen of Age Concern, Vera Coppard of Travel Companions, Margaret Ward of the University of the Third Age, Sally Finnis and my editor, Gillian Clarke.

I am also indebted to Jim Bennett and Adrian Drew of Tripscope; Dr Anthea Goode; Sally Hamilton of the *Daily Express*; Julia March for researching Chapter 8; PC Elizabeth Hand of the Crime Prevention Office of the Paddington Division of the Metropolitan Police Force; Rebecca Rees of the Automobile Association; Leonora Curry Barden and Trudi Thompson who gave me many valuable insights; and the people who were interviewed for case studies. Some names and identities have been changed.

1.
Decision time

To travel is to discover – yourself and new places, experiences and sensations. Whether you are single, divorced or widowed, the delights of travelling are accessible to everyone these days. One of the bonuses of getting older is that there is more time to explore the world – whether it's because you can now enjoy the luxury of retirement or simply have more time to spend now that career and family are no longer so all-consuming. You can decide what sort of travel you wish to engage in – a package holiday, an independent trip or a mixture.

At one end of the traveller spectrum is the intrepid backpacker who dares to go where few have gone before; at the other end is the tourist who just wants to relax and unwind in pleasant, sunny surroundings for a couple of weeks. Most people probably fall somewhere in between these extremes, although, after retirement, having a holiday for relaxation purposes can take something of a back seat as many are looking for stimulus and the chance to enter a new period of growth and fulfilment.

There are many advantages of travelling at a later stage in life – not least being the opportunity to make use of the off-peak season when airlines, ships and hotel operators adjust their prices downwards. Not being tied to going away only in the school holidays any more, for instance, means that some countries can be experienced at their best and most beautiful times – many Mediterranean countries are at their most idyllic in spring around the month of May, before the scorching sun shrivels all but the hardiest vegetation.

First things first

You don't have to battle with the elements or rough it in a tent on the side of a mountain to travel independently these days, although there are a few hardy souls for whom this is the ideal. There are more travel opportunities than ever before and the buying power of the older traveller is really beginning to come into its own. Whether you are a fit older person who would like to go trekking in Nepal or someone with a few health problems that mean you'd prefer to settle for a reviving sojourn in one of Europe's many spa towns, there are options to suit most pockets.

The secret to successful travelling is to recognise both your limitations and your capabilities. But first decide where you want to go and what you want to do. Most people travel for a mixture of reasons: to enjoy and learn about new places, to revisit old haunts, to relax in the sun or to recuperate after an illness or operation. You may have a few places in mind that you've always wanted to visit, or you may be approaching your trip with a blank slate.

Before you decide anything – or ill advisedly take the first offer that comes your way – ask yourself the following questions:

CHECKLIST

- Do I want to go abroad to a new place? Would I prefer to visit a country I've been to before? Or would I prefer to visit somewhere in my own country?
- Are there any cultures or places that particularly interest me?
- Are there any physical reasons or illnesses that might influence my choice of destination?
- Is going to a country where I can speak the language an important consideration?
- What type of living conditions/level of comfort am I looking for?

- Is the weather important? Do I want some sun/snow?
- Does the thought of fitting in with unfamiliar ways and customs worry me? Do I see it as a challenge?
- How active or leisurely do I want to be?
- Am I limited by time?
- How much money can I afford to spend?

Take as much time as you need, making a few notes if necessary, to answer these questions fully. Then draw up a list of the options available, and from there you can begin to consider possibilities. But before you commit yourself to any course of action, consider whether you want to travel alone, with a companion or in a group: this is the subject of the next chapter.

A number of helpful organisations are mentioned throughout the book. Their details will be found in the 'Useful addresses' section at the back, grouped by category, so you can browse through them or look up specific topics.

2.
Travelling alone, with a companion or in a group?

How you adapt to travelling and holidaying alone or with other people will naturally be partly determined by your own personality and circumstances. Both options are equally challenging in different ways. Travelling solo throws you back on your own resources, whereas the demands exerted by a companion or group mean that you have to try to maintain a balance – between making the most of the companionship on offer and making sure you get what you want out of the holiday.

Going alone

Women who have travelled alone almost always say how much good it has done their self-confidence and how much stronger they feel. Their experiences have made them realise that they can cope on their own in a variety of different circumstances and they find it tremendously liberating. The advantages of being a single traveller are that you get to know places really well – you really do see more than if you're with a companion or a group. On the other hand, you will have to be self-reliant and occupy or entertain yourself for much of the time – and this is not to everyone's taste.

Some people, of course, are alone but not by choice. One of the immediate consequences of bereavement or divorce can be an enormous loss of self-confidence, and it's perfectly understandable to feel cautious and anxious about travelling in these circumstances. If you are still feeling vulnerable, the thought of a holiday alone in the UK can fill you with dread, let alone travelling to some far-flung corner of the globe.

While your feelings of grief over bereavement or separation are very intense, it is probably best not to go away alone or with a group of completely new people – although for some a spiritual retreat (see Chapter 7) may be beneficial. Perhaps a quiet relaxing trip with a friend or relative to one of Britain's many beauty spots is a possibility. There will be time for other holidays at a later stage when you have worked through more of your grief and your self-confidence is returning.

Half-way house

A good intermediate way of combining independent travel with the security of knowing who you'll be staying with is to travel by yourself but to stay with a family, either when you reach your desired destination or from time to time in the course of your journey. The organisation called Women-Welcome-Women (see p 122) can put you in touch with women from 64 countries world-wide (from Australia to Zimbabwe), many of whom are only too happy to have a guest to stay – on the understanding that if they come to your country you will provide equal hospitality.

Whatever your personal circumstances, before you start to make any plans ask yourself: 'Will I truly enjoy being on my own?' If you unequivocally answer 'No' – for whatever reason – then reconsider your options. There is absolutely no point in travelling alone if it is going to make you miserable.

If you're undecided – and it is often difficult to imagine how you'll cope, particularly if you've always been away with friends or a partner – you could always test the waters by spending a weekend away somewhere in the UK where you've never been before. The Automobile Association's *Bed and Breakfast Guide* (£7.99, updated annually) has a good range of 'B & Bs' to suit most pockets, so your experiment shouldn't set you back too much even if you decide that travelling solo is not for you!

Another way to help you make up your mind is to go on one of the Marco Polo Travel Advisory Service's Women and Travel seminars (see p 121). Aimed at the independent traveller, the day's seminar gives an excellent, no-nonsense introduction to the practical aspects of travelling, with sessions ranging from what to take with you to self-defence. You'll also meet people who have travelled widely.

Enjoying your own company

The secret of enjoying holidaying alone is to cultivate the art of liking your own company. This means using the time profitably to explore and develop sides of your character that perhaps, owing to pressures of life at home or at work, you simply haven't been able to concentrate on before.

CASE STUDY *'I was ready for anything when I returned'*

Rose Jackson, aged 55, went away for the first time by herself to recuperate some months after her divorce became final.

'Since it was my first time away ever on my own, I didn't want to go anywhere that I'd have to speak a foreign language; I've never been particularly good at them. And I wanted a place that held no memories – I didn't want to go anywhere that would remind me of my ex-husband and the happy, early days of our marriage.

'Sun and sand holidays have never been my thing, but I love looking at the sea so I decided to book a break on Jersey. After an uneventful flight where I spoke to no one and wondered whether my holiday was a good idea, I arrived at my comfortable, small hotel with a room with the proverbially lovely view.

'Sitting on the balcony before dinner, I thought to myself, "If I'm just going to sit around it will be awful", so I made it my aim to find out and experience as much of Jersey as I could within a week.

'Next morning I went to the tourist office and collected a mound of brochures and decided to hire a car, although I hadn't done much driving for the last few years. I was a bit nervous, but I stopped worrying about grinding the gears after a few miles. If I'm honest, I didn't expect to enjoy myself very much, but I had a superb time.

'I wasn't there to make friends – more to think out my life and what I was going to do next – so I didn't seek people out, but my travels took me to attractions ranging from a stud for Shire horses to a lavender farm. Coming back in the evening and sitting on my sun-lit balcony before dinner, I found my problems diminishing with every day.

'The week went all too quickly. On the flight home I felt the most enormous sense of accomplishment and funnily enough I found myself chatting easily to a woman across the aisle. We exchanged numbers and arranged to meet. I was a different person from the one who had flown out a week earlier, deep in her own thoughts and worries. I wouldn't necessarily go away alone every time – but I was ready for anything after this particular break.'

Creature comforts

Being alone gives you the chance to set aside extra time to pamper yourself. If you're feeling tense and worried about life – or even about the holiday itself – take a relaxation tape and small tape-recorder with you. Half an hour before you go to bed each night play it and follow the instructions: you'll be amazed at the difference it makes to your physical bearing as well as to your state of mind.

Other aids to relaxation include aromatherapy oils such as geranium, sandalwood or rose. They shouldn't cost more than a few pounds per bottle, take up very little space in your baggage and last for a long time. When you have a bath, sprinkle five or so drops into a teaspoon of milk (so that the oil does not float on top of the water) and add to the bath water, which should be warm but not boiling. Relax for ten minutes or so in the bath – and inhale the pleasant aroma. Travelling, after all, should be fun and restore your sense of wonder in the world; it should not be a marathon race, which can happen all too easily if you're not careful.

Useful traveller's 'friends' include a small blow-up cushion (which folds away into next to nothing) to place behind your neck on long coach or train journeys and a small bottle of eau de cologne or your favourite scent to dab on the pulse points (behind the ear and on the wrist). Of course, these small additions to your luggage could just as well be taken if you're travelling with a companion or in a group – though when you're travelling alone, little luxuries somehow are appreciated more. Treat yourself to a glass of sparkling wine with your evening meal, a tip recommended by Elizabeth Green, aged 62, who prefers to travel alone because she 'can really get to see how people live and operate'. As she says, it won't inebriate you but it gives your holiday a bit of style.

With all the new sights and experiences, however, it's important that you do not ignore the need to rest. Traveller's fatigue is more common than most people think – and is particularly likely to be experienced by the single traveller who has set herself too tough a schedule. Even sitting on a train all day can be surprisingly tiring, as are delays of any sort.

If you are either travelling or busy with intensive sightseeing, try to take at least half a day a week to relax and unwind, catching up on sending postcards and doing the washing. Take a whole day if you feel you need it. Other useful suggestions to help combat fatigue include asking a steward to carry your hand luggage on and off a plane – many are only too pleased to help (Elizabeth Green) – or arranging in advance with the airport to provide a chair or buggy to take you the often enormous distance to your gate for take off (Leonora Curry Barden).

Above all, you should not allow yourself to get to the stage where you are just keeping up with your itinerary, at the expense of all else.

Expand your horizons

Travelling or holidaying by yourself also presents an excellent opportunity to further your hobbies and interests. Take advantage of the new perspective that your travels will give you by trying one or more of the possibilities outlined below. Any equipment needed does not weigh much or take up much space in your suitcase. You may have others you wish to add to the list, or you might decide to link several to, say, have photos and a written record of your experiences, which could be pasted into a photograph album when you get back. Or, as many travellers find, these become the basis for a talk to give to friends and family, or organisations of which they are members.

- **Catch up on your reading**. Is there a particular author or subject you wish to know more about? Or do you want to branch off in a new literary direction – perhaps read some of the novels from the country you are going to? Most major authors are readily available in translation these days – your local bookshop will be able to advise.

- **Keep a diary of your holiday or trip**. If you don't feel like writing everything down, or have difficulty doing so, a small portable tape-recorder can be used to record your impressions and thoughts.

- **Draw or paint**. This does not necessarily involve carting around a vast amount of equipment. A set of pastels and/or pencils and a drawing pad, or a small set of watercolours, need not take up too much space.

- **Revive the dying art of letter writing**. Write letters (or short stories if the spirit takes you). Send a long, descriptive letter to your children or grandchildren, telling them what the place you're visiting is really like. Alternatively, you could record this on a cassette tape.

- **Make a visual record of your holiday or trip with photos**. If you are going on a long journey, take a tip from a seasoned traveller, Jane Copeland – who went round the world alone on a epic trip lasting more than four years after she retired – for a quick and easy way to cope with reels of undeveloped film. Rather than lugging around a fast-growing collection of films ready for developing, send them back by post to your home address from time to time. That way, you won't have to worry about losing them, and if you enclose a brief note saying what they are pictures of, you'll know instantly what they are.

As a point of courtesy, always ask if you may photograph anybody before you do so. In Muslim countries the answer will be 'no' because their religious beliefs forbid reproducing the human form.

Solutions to some common worries

Apart from considerations of safety, most of which can be dealt with by taking sensible precautions (see Chapter 5), the main worries that many single travellers have are largely social ones – which is only natural if you're shy by nature or haven't been on holiday by yourself for many years.

Going into a restaurant

Going into a restaurant alone can be quite nerve-racking. Where do you look? Will you get the worst table by the kitchen? Will you be served?

Rose Jackson (see the Case study on p 141) found that a good solution to the restaurant problem was to eat fairly early in the evening before all the parties and couples came in. 'I felt a bit awkward to start with, but I got extremely good service from the waiters, as they weren't yet working at full tilt, and I didn't feel that I stuck out like a sore thumb,' she says.

If you're feeling self-conscious, you can always fall back on the time-honoured way of eating alone in a restaurant – taking a book or newspaper along with you. If you don't fancy eating your meal with your nose in a publication, there's nothing to stop you approaching another solo diner. Use common sense here; if you seat yourself at a table near the person, it's usually fairly easy to open a conversation and see whether they are in need of company. From time to time, I find I get talking to someone in the hotel lobby while I'm looking through brochures or details of excursions, and on one or two occasions I've found it easy to suggest that we dine together.

Meeting people

Meeting people, whether from the local population or of your own nationality, is much easier than you think. Take advantage of being on your own to mingle with other people in the daytime and find out about your host country. Initiating conversations is not as fearsome as it seems – asking directions is an easy option for beginning a conversation, as is the useful old standby of 'Where are you from?' or 'Where are you going to?'

Wherever you go abroad – even if English is widely spoken – it's a good idea to learn the basics in a foreign language. However stumbling your attempts, many nationalities will often respond favourably to this gesture of good will, making it easier to meet people. Learn at least to say, 'Please', 'Thank you', 'Yes', 'No', 'Do you speak English?' and the numbers you need for asking about prices, telling the time, paying the bill in restaurants and shopping for souvenirs. Linguaphone courses can be borrowed from many local libraries and the Teach Yourself series of books includes all the popular foreign languages, and quite a few minority ones as well. Many adult education establishments hold a variety of foreign language classes at various levels. Who knows, learning a language may even become a new hobby as a result of your holiday.

Finding company

If you get to the stage in your travels where you need company and aren't finding it, make for a tourist area where you will meet other travellers. Or take an organised excursion, such as a boat trip, where you can meet others in the same situation as yourself. Another possibility is to go to an English (or American) church service – there are many in cities and resorts abroad – even if you are not a regular church-goer. The English church is often a meeting place for many of the British community, from Embassy staff to expatriate business people and their families.

Two's company

Travelling with someone else is a good way of sharing the whole experience – and it can be good to have someone to compare notes with. If you're feeling tired or worried, you will have someone to turn to, and someone with whom to share the costs. If you decide to share a room, you'll avoid single room supplements.

On the down side, there is the possible problem of constantly having to compromise in order to get on reasonably well with someone. Combining two different points of view on what to see and having to accommodate someone else's habits can be frustrating at times. Coupled with the fatigue that journeys may cause, tempers can sometimes flare at the end of a long day.

Finding a companion

It can be difficult to find someone to go away with: most people you know are part of a couple, or your only single friend prefers an art appreciation course in Florence to the wine-tasting tour you want to go on in France. One way to find a companion is to contact one of the agencies that puts people in touch with one another. Travel Companions and Single Again are two organisations (see p 122) that can put travellers of a similar age and interests in touch with one another.

CASE STUDY *'Half the fun of a trip is comparing notes'*

A psychotherapist by profession, Mary Black has been a widow for seven years now. Aged 72, she increasingly found that 'everyone I knew had a husband or sister or friend to go abroad with'. She contacted Travel Companions, who duly put her in touch with three other people.

'I'm not the sort of person who'd be happy holidaying on her own. I don't mind travelling alone to meet friends or relatives who live abroad but I like company. For me half the fun of a trip is to compare notes with someone else. And now that I'm widowed, I found increasingly that everyone else seemed to have a husband or sister to go away with.

'One of the people I contacted sounded really nice on the telephone and we quickly arranged to meet for a day in Brighton – she lived quite a long way away. We sat on the pier and just talked and talked. What quickly become apparent was that we had a lot in common: she was a nurse and so, like me, in one of the caring professions, and a widow as well.

'We decided to meet again, just to be sure, so she came to my home. I very quickly had no doubt that she had two of the qualities I value most highly – she was kind and had a great sense of humour. That decided it and we set about arranging a holiday together.

'Some compromise was obviously necessary – we had to pick somewhere that suited our respective finances – but without too much trouble we settled on a coach tour of middle and northern Italy.

'We got on tremendously well throughout the whole holiday, which was a good thing as most of the people on this particular tour happened to be couples. We had a lot of fun; particular highlights I remember are walking in the Italian countryside and a wonderful trip to Venice.

'I think the secret of choosing a companion is to meet up with them several times before you decide to go away, to make sure that you have enough in common. I'd gladly find another travel companion this way – for me it was the perfect solution.'

Hints for companionship

Choosing a companion is largely a question of common sense, and following your instincts as to whether you trust and like someone. But having to compromise with someone you don't know well – perhaps even a complete stranger – requires tact and flexibility. You don't have to agree on absolutely everything to be able to enjoy a perfectly good holiday together, but it helps if you've got the basics sorted out before you set off.

- Take time to get to know one another well before you go. This may mean meeting up on several occasions – perhaps going out for a meal or a day trip to 'break the ice' and ensure that you feel you have enough points of common interest to make the trip mutually enjoyable.

- Discuss what each of you aims to get out of the trip. It will help if your intentions are broadly similar – you don't want a situation where one person wants only to relax while the other is determined to see all the art galleries.

- Make sure you settle any money questions before you go. Discuss your budget in detail – do you have the same amount of money to spend? Decide whether you'll have a kitty for joint expenses – so much easier if you're eating out often, than having to divide up the bill each time. But beware the friction that can arise if one of you feels that the other isn't paying her fair share because she has three expensive courses and wine to the other person's two courses and mineral water. Agree on how much you're going to set aside for emergencies and, above all, be honest about your attitudes to money.

- Talk through some of the problems that might crop up – delays, difficulties with hotel bookings and so on – and work out a strategy for solving them. That way, if things do go wrong (and no journey is without a hiccup of some type, however minor), you can solve the problem together immediately and not waste time having to come to an agreement as to how to proceed.

- Discuss and agree how much privacy each of you will need. Are you both happy with sharing a room? Will either or both of you want a little 'private' time each day to take stock and relax alone? Will there be times when you might like to spend the whole day apart – doing different activities? Agree before you go, if necessary, what activities could be done separately.

Plan your own trip with a friend

Of course, you could always plan your own trip with a friend. Within Europe there are city breaks to fascinating cities such as Amsterdam, Prague or Munich, to name but three. A useful trick to help you orientate yourself quickly in a strange city is to go on a guided tour in a coach as soon as is convenient after your arrival. You'll get a reasonable overview of where the main sights are and, more importantly, whether they are within walking distance. Then you can plan your sightseeing more effectively.

For a European city break, taking a plane will get you there in a few hours. If you wish, search the national and local papers for discounted air deals – but book only from an ATOL-holding agent. An ATOL, or Air Travel Organiser's Licence, is required by law for anyone selling charter flights or holidays to the public: it provides protection to the consumer through a fully backed up bond guarantee. You can tell instantly if a company is an ATOL holder by the registered number that appears on its logo.

If you wish to plan a long-haul trip – and more and more retired people are setting off around the world – a number of organisations can assist. Trailfinders Travel Centre, for instance, has a string of consultants to help you book and arrange your itinerary.

The group

If you haven't done much travelling, going in a group is an excellent way to start, as everything is organised for you and there's always a sense of security. Indeed, it's the still only way to get to see some countries or remote places where few single travellers would dream of venturing.

A group tour can also be good for meeting people, although if you're single you'll want to avoid trips where there is a preponderance of couples. Apart from singles holidays, good ways to meet other single travellers are often special-interest groups and activity holidays.

Information on interesting and unusual tours for single people may be obtained from good independent travel agents. AITO (the Association of Independent Tour Operators) has a list of overland tours and group travel, as does Trailfinders, which has branches in Birmingham, Bristol, Glasgow, London and Manchester.

Other good places to find a group include educational and study organisations such as the University of the Third Age (U3A) which has a Travel Network through which members organise special-interest study tours and house parties for small groups in the UK and abroad. Some of its recent tours include New England in the autumn, and the seldom-visited Yunan in southwest China, just off the Burma Road, where most of Britain's camellias, rhododendrons and primulas originated, as well as trips nearer home such as gentle walking in the Lake District.

Or try the Association of Cultural Exchange, which is responsible for a wide range of world-wide study tours covering such subjects as natural history, music and literature. The biggest holiday organisation for the over-50s, Saga Holidays, also has an enormous number of package tours and special-interest holidays.

In a nutshell, the disadvantages of group travel are that the individual's freedom can be lost, organisational matters can take up valuable time, and sometimes the whole schedule can seem inflexible. You may also find that group activities tend to hinder you from experiencing first hand the culture that you are visiting, as tour members often become so bound up with group activities and social events that there isn't time to make much contact with the local people or see how they live.

The best way to make the most of a group is to sift through the options available to you until you find one that most covers your needs:

- Look at all briefing material that is sent to you. Does it represent the country/or special interest in a way you find appealing?
- Do the group members have any say in what happens on the tour?
- Are you allowed any time on your own for individual activities/sightseeing?
- Who is running the group? Do they have the necessary qualifications (particularly relevant for special-interest activities and study tours)?

- How large is the group? Sometimes a smaller one can be friendlier and you may have more say in its running.
- What is the ratio of men to women (if it is a mixed group)? Many organisations now try to see that the ratio is about half and half or at most a maximum of 60:40 (women to men) in mixed single groups. This can't be guaranteed but it's always worth asking – particularly if you like a 'balanced mix'. As one tour organiser says, 'It's a fact of life that more older single women than men tend to go on these trips.'
- Don't be afraid to double-check that the people going on the trip are – broadly speaking – of your age range and type. For example, it's no good going on holiday to see the treasures of ancient Greece with a group of people who really only want to relax and eat and drink.

Starting your own group

If you can gather enough like-minded people there's no reason why you shouldn't form your own group and go on a package holiday or a group tour. A good place to canvass for fellow members would be at, say, your evening class or local club or society.

If you can gather 15 or more people together, you can usually qualify for group discounts on both your accommodation and your travel. Discounts may be a reduction of up to 10 per cent or, more commonly, one person in every 15 going free of charge. These concessions vary from operator to operator, so you'll have to approach several to find the best price.

This is a good way to save a small proportion of the costs, but, as any holiday company can tell you, reconciling everybody's interests is not an easy – or relaxing – task.

Making the most of a group

Groups offer instant companionship and, more often than not, a choice of companions. Some people swear by them as a way to meet new people; others go along mainly to further an interest, whether it's archaeology or gourmet cookery, and are pleased to meet like minds who share their enjoyment of a subject.

CASE STUDY *'You don't have to think about day-to-day things'*

Thelma Stubbs, aged 71 and a widow, has been on more than ten group tours in the last ten years or so, and has travelled to countries such as China, India, South Africa and Egypt.

'What I like about the guided tour is that you don't have to think about day-to-day things. You can concentrate on seeing the Taj Mahal, or whatever, without having to worry about connecting flights and the like.

'I've saved my pennies over the years and been on some fascinating journeys. Often tour leaders have been experts in their field and I've found it all very stimulating. I've always felt safe and I think that in itself enables me to enjoy the trip more.

'I would recommend taking the bed and breakfast option, if you are offered it, as that enables you to eat out at a restaurant of your own choosing in the evening, rather than being committed to hotel fare.

'I must say, now that I'm older I do enjoy the comfort of a good hotel, and I probably tend to like travelling with smaller groups as they can often adapt the pace to suit you. When I was in China last year, for instance, if some of us felt a bit tired during sightseeing, the tour leader used to arrange a break for a while, and then we'd set off again.

'The only criticism I would make of some tours – maybe I've been lucky but I really haven't come across any big problems – is that occasionally one has to catch connecting flights at unearthly hours in the morning.'

Solutions to group problems

Feeling ill at ease in a group is a problem suffered by many, whether through shyness or a lack of confidence. At some time or other most people have probably found themselves in a social situation where the conversation has dried up and been replaced by an embarrassing silence. It happens to the best of people, usually at the worst of times, such as when they are in an unfamiliar social setting.

If you're feeling nervous or ill at ease – which might well occur if you are in a situation where everyone else seems to be in couples or groups, or at least know each other – take heed of author Don Gabor's tips to encourage conversation. In his book *How to Start a Conversation and Make Friends* he recommends: 'Be receptive. One of our most important conversational skills doesn't come from our tongue, but from our body. Research has shown that more than 70 per cent of communication is non-verbal. Our body language often communicates our feelings and attitudes before we speak.' What he means is that people are less likely to approach someone who stands with arms folded or legs crossed (both defensive postures) or with a hand covering their mouth (another 'stay away' signal). So smile, and stand in a relaxed open posture that will indicate you are receptive to conversation.

Other tips that psychologists recommend include:

- Search out people with receptive postures in the room. Knowing that they are willing to be approached will make it easier for you to go up to them.

- Once you have been introduced to someone or made the first approach, try not to worry too much, as many anxious people do, what you are going to say next. Concentrate on listening to what the other person is saying: the art of conversation is as much about listening as it is about speaking. As you become more proficient, you'll be able to think up many ways of continuing conversations, although take care never become too enthusiastic by allowing a chat to become a cross-examination. Remember, too, that there may be other shy people in the group and you can help each other to relax and chat together.

- Avoid the traditional taboo subjects of death, gory crimes, politics, religion, unhappy events and personal gossip, when you meet people. Avoid, too, getting things off your chest and 'unloading' as it only creates a very negative first impression.

- When you meet someone new, don't think you automatically have to be the life and soul of that particular conversation. The vital trick is to keep drawing the other person into the conversation with questions about his or her life and opinions. Focus your attention on them; then you can build up a discussion together.

Having second thoughts about travelling?

Don't worry, it's perfectly natural, particularly if you haven't done much travelling before. You might be worried about what can go wrong while you're away, or even what can go wrong at home while you're not there. Provided you have made good back-up arrangements at home (see Chapter 4) and not left things to chance, you have little to worry about. Leave full details with a friend or relative of where you may be contacted at all times during your trip. If you're going abroad, take comfort in the fact that it's easy to ring home from most countries these days. The cheapest way is not to use the hotel phone – some can cost up to five times as much – but instead use a public call box. International Direct Dialling is the fastest, cheapest way to place a call; in many countries you can now buy slot-in phone cards so you don't have to keep feeding a public phone with coins or tokens. As a last resort, you can always make reversed-charge calls without spending any cash, but they are costly.

A final word to the anxious and worried who, quite understandably, may be feeling more daunted by the minute at the thought of venturing away from home. Remember events from earlier in your life – the things you dreaded doing and the feeling of accomplishment once you had done them: when you had braved your first job, or coped with a domestic disaster. The world outside is not as difficult or dangerous as you might think: the trick is not to step too far at first without visible means of support from a friend or a group.

3.
Travellers with special needs

Whether you suffer from a condition such as arthritis or osteoporosis (brittle bones) that has developed gradually as you have got older or from a disability that requires the constant use of a wheelchair or guide dog, there are now more opportunities than ever before to travel at home and abroad.

The amount of suitable accommodation has increased enormously. In the UK, for example, the English Tourist Board now lists more than 400 places to stay that have been graded according to three levels of accessibility for wheelchair users. Category 1 accommodation is accessible to all wheelchair users, including those travelling independently; Category 2 is for those who need assistance; Category 3 is for wheelchair users able to walk short distances.

The development of amenities and equipment available abroad has been uneven, though, and there is by no means enough of them; for instance, in many countries it is still difficult to hire a car with hand controls.

On the whole, the USA, Canada, Australia, New Zealand and countries in northern Europe are the most geared up to accommodate travellers with a disability. Nevertheless, depending on what your needs are there is no reason why, with good planning and determination, you should not be able to travel to many other destinations as well. Some disabled travellers have ventured as far afield as India or even China independently; others have made the most of the many tours (both commercial and charity-run) to take holidays ranging from short special-interest breaks such as painting or drawing to venturing on to the ski slopes.

CASE STUDY *'Do your research first'*

Christine Warburton, 51, has suffered from rheumatoid arthritis since she was very young and her mobility is severely limited. She is able to walk a few paces on crutches but cannot bend her hips or knees. However, she travels regularly and has fulfilled her greatest dream – to go to Venice.

'I've always had the travel bug. Despite some difficult journeys – with every delay and problem under the sun, when I've thought "never again" – the memory of the horrors soon fades and you start to prepare for the next trip.

'Although I always travel with someone, for me the main worry is being out of control. How will I cope if my arthritis flares up? How will I manage without my special equipment, such as the hoist which now helps me into my bed? Without a doubt the facilities and access to places for disabled people have improved no end – 20 years ago even going out in a wheelchair was a nightmare. It's not perfect now but things are improving.

'The important thing, I've learned, is to keep a sense of humour and go to places where they can cope with your disability. Do your research first and reassure yourself: many countries nowadays have excellent medical facilities, so there's no need to be concerned on that score.

'My main problem is not being able to bend my hips and knees, which makes airplane journeys uncomfortable, to say the least, as I often end up travelling on the seat at the front by the bulkhead. Anything more than a few hours in the air quickly becomes very uncomfortable, so long-haul flights are out of the question, but it hasn't stopped me travelling to places such as Italy, Cyprus, Malta, Austria and Switzerland.

'Be realistic about your capabilities – don't say, for instance, that you don't need a wheelchair if you do – as on holiday you'll spend much more time out and about than at home. But I can promise you that the sense of trepidation we all feel at times like this will be more than outweighed by the sense of adventure and achievement when you're there.'

For some, however, there are other problems to surmount, not least when their disability is in effect invisible to the outside world: people with poor sight or hearing loss, or perhaps both.

CASE STUDY *'Pick a structured holiday'*

A teacher by profession, Careen Bradbury, 60, is registered blind, although she has some useful residual vision. She has travelled widely since her late thirties and in recent years has taken to going on activity holidays with a totally blind friend.

'In my case, the difficulties are more subtle because it's a hidden handicap. In some ways it's not daunting to travel abroad, as Brighton is just as strange as Benidorm for me. But I must admit I wouldn't have gone on my first trip – by coach through Europe with a tour company – had it not been for the encouragement of a friend. I simply didn't think I would enjoy travelling alone.

'But there's been no stopping me since. The next summer I went on my own on a coach tour to Italy, Yugoslavia and Austria; trips with companions to cities such as Paris and Venice and to countries such as Norway and other European destinations followed. This autumn I intend to cross Canada alone by train.

'If you have a visual handicap, the secret is to pick a structured holiday; for example, travelling by train or bus with a courier who will smooth the path for you. And make sure you make as many arrangements in advance as you can. Getting to grips with the itinerary before you go will help you decide what you want to visit on your trip. Knowing a foreign language can be a great asset, too. By helping your fellow travellers with, say, ordering a meal in the language, you will be contributing to their greater enjoyment and they in return will be only too happy to assist you with a friendly arm when necessary.

'Of course, there have been hitches and problems, but I've found people unfailingly helpful wherever I've been. I wouldn't have missed meeting so many new people and enjoying experiences such as skiing cross-country.'

Pre-trip tips

It's your decision as to whether you travel in a group or independently (perhaps with a companion), although group travel may be preferable if you need physical assistance, because volunteers are readily available to help when necessary. For inspiration and ideas on where to go read the Rough Guide's *Nothing Ventured – disabled people travel the world*, which also contains useful travel notes at the end of each chapter and a realistic appraisal of the varying facilities, access and transport systems offered world-wide.

The Royal Association for Disability and Rehabilitation (RADAR) publishes a guide called *Holidays and Travel Abroad* that is updated annually and contains information on more than 110 countries world-wide. RADAR also publishes *Holidays in the British Isles*, which details accommodation ranging from camp sites to hotels and special-interest holidays such as riding and fishing. Contact, too, the tourist office of the country you wish to visit: many of them have information about facilities for disabled travellers (see Chapter 8).

Once you've decided where you want to go, you can start planning the whole trip. The secret of successful planning is to try to think your way through the trip, anticipating and making contingency plans for your particular 'trouble spots' – which are often travel-related – before they occur. Of course, there are bound to be hitches and hold-ups, no journey is without them, but at least you'll be prepared and this should allay many of your worries.

Find out as much as possible about the place you wish to visit, including details of the terrain if you have problems with mobility. Ask questions such as:

- What are the amenities like – are shops, entertainments, beaches accessible?
- Are the facilities suitable for my particular disability, such as menus and hotel safety instructions in braille?
- Can my special diet be catered for?

Sometimes guides detailing access to some countries' facilities are available – again it's worth consulting the national tourist office of the relevant country to find out. Some travel agents can obtain printouts listing the facilities available at resorts.

There are two useful organisations you can contact for advice and information. One is Tripscope, which offers free advice on travel in the UK and abroad for disabled and older travellers. The other, Holiday Care Service, offers a wide-ranging service from details of accessible transport to holidays. They can also help you with reservations.

If you do not wish – or are unable – to travel alone and a companion is not readily available, Holiday Care Service has a factsheet listing organisations that provide escort services; RADAR publishes a list of care attendant agencies. Try to spend some time with the prospective helper before you go away so that you can establish whether you'll get on, and make sure you agree on the finances and payment before your trip starts.

Medical requirements

If you're going far afield, a full medical check-up is always a good idea, as is finding out where the local hospitals are located and any specialist treatment can be obtained. Check with your doctor which immunisations you will need. British Airways Travel Clinics can also advise on immunisations, or your disability organisation may be able to provide the necessary information (see Chapter 6 for general information about immunisations). Some conditions cannot tolerate certain immunisations that use live vaccines (some severe types of arthritis needing high-dose steroid treatment, for example). This may alter your travel plans, so it's best to look into this early on rather than be disappointed once you've set your heart on going somewhere. You may also need permission from certain countries (such as Greece) to take some medications in; contact the embassy of the country you're going to or the Home Office.

Be sure to take more than enough of all your medicines, and a written prescription – with dosages clearly stated. Depending on where you are going, you might get the prescription translated before you leave (see Chapter 6 for other medical tips). Like all travellers, you should always carry your medication in your hand luggage; it's too important to risk losing it.

For extra peace of mind, some disabled travellers take a doctor's letter with them explaining their condition and giving details of their treatment. (This is particularly useful if you use a hypodermic syringe, because some customs officers become very suspicious of such an item in your luggage.) You could always get this translated if necessary. (See Chapter 6 for general details of a first aid kit to take with you.)

Insurance

Unfortunately, even wheelchairs have been known to be stolen, so it's essential that you are adequately insured for both yourself – this means that the policy should not exclude people with a pre-existing medical condition – and your equipment and medicines. There are a number of specialist policies designed specifically for disabled travellers: they are more expensive than general policies but in return may offer extra benefits such as special cover for wheelchairs or for dialysis equipment or repatriation by air ambulance. Tripscope and Holiday Care Service can advise on organisations to contact. (See also Chapter 4 for general information on insurance.)

Booking your trip

You'll have a smoother, better trip if, when booking, you consider the following points.

- Outline clearly and honestly to the travel agent/tour operator what disability(ies) you suffer from, including any additional difficulties, such as incontinence or problems with eating. It's essential to do this for insurance purposes, as withholding such information could affect any claim you might make while travelling.

- Explain clearly what facilities you are looking for at your destination: for example, ramped or level access to accommodation; a lift that is large enough to accommodate a wheelchair user and possibly a companion; suitable access to bathroom; access to local shops (remember to give the measurements of your wheelchair in centimetres).
- Point out the places where you will need help on the journey, such as travelling to the airport/port and assistance on boarding trains/airplanes/ferries.
- Arrange for a medical certificate, if you think you will need one. Few disabled travellers do, in fact, need to prove that they are fit to travel, but they will need to fill in an INCAD (Incapacitated Passengers Handling Advice) form for the airline, which gives details of any special needs and assistance required. If you have a medical condition, the second part of this form, the MEDIF (Medical Information Form) will have to be completed by your doctor, certifying that you are fit to travel. If you have a stable medical condition, obtaining a Frequent Travellers' Medical Card will save your having to fill out a form each time you go abroad.

Before you go

It's essential to make a checklist of any extra equipment you think you will need – for example, a wheelchair repair kit or commode. If you are travelling within the UK, items such as manual wheelchairs, commodes and back rests can be borrowed from the local Red Cross, saving the problems of transportation. Most of these are free but some branches make a nominal charge. There are more than 1000 medical loan depots throughout the UK and you should contact your local branch of the Red Cross for details. If you wish to hire an electric wheelchair or three-wheeled scooter from a place near to your destination, Tripscope or the Disabled Living Centres Council can provide you with details.

Other special aids that may be worth investing in include a collapsible commode, for times when it's difficult to get to the loo, a set of bed blocks for raising a low bed and a collapsible stool. The Disabled Living Foundation can advise.

Main travel options

Air

Most airlines these days provide a good service when it comes to helping passengers on and off aeroplanes, provided full details of your needs are supplied when making the booking. Some airlines have more experience than others of carrying passengers with special needs, but facilities vary greatly on board and you would do well to speak to friends with similar disabilities and the relevant disability organisation before booking. And, of course, shop around for the best value in fares.

Generally, powered wheelchairs and dialysis equipment are carried free world-wide, and guide dogs may accompany passengers on flights within the UK. If you cannot attend to your personal needs yourself, most airlines prefer that you travel with a companion.

Once in the air, two problems that may need to be overcome are the aircraft seat and the toilet facilities. On the whole, aircraft seating leaves much to be desired – in economy class, you're unlikely to enjoy a comfortable seat on a long-haul flight – as do the narrow aisles and tiny toilets which allow very little movement. Ask the airline when you book if you can be seated where there's extra leg room; they will often try to help, although this can't be guaranteed.

If you anticipate difficulties using the lavatory, talk to your doctor about the various aids available. Some airlines carry aisle wheelchairs, which will help in the trip to the loo at least, but it must be said that the main advantage of air travel is being able to get to your destination quickly rather than especially enjoying the journey.

Helpful hints

If you live a long way from the airport, rather than arrive exhausted for your flight, spend the night before at a nearby hotel. Breaking your journey in this way will eat into your holiday time, but may make your journey a pleasanter affair. Heathrow Airport produces a free guide to its facilities, *Traveller's Information Special Needs*, which can be obtained by ringing 0181-745 7495.

The Disabled Living Foundation produces a booklet, *Flying High* (£2.50 including p&p), which gives practical information about flights. The Air Transport Users Committee produces *Flightplan*, which has a section for the disabled. Blind travellers can obtain *Plane Easy*, a tape available free from the Royal National Institute for the Blind.

If you would like more information about your flight and boarding procedures, do not hesitate to ask for this when you are making the booking.

Rail

Generally speaking, travelling by rail requires a great deal of patience, for the standards of rail services in Europe are improving only gradually. There are still far too many trains with narrow doors, narrow aisles, carriages that can be boarded only by steps, and too many stations with gaps between platform and train. Of the European countries, Norway currently offers the best train service for the disabled traveller: its trains have hydraulic lifts, accessible loos and specially adapted coaches for wheelchairs. Switzerland also has a good service.

Handy hints

If you are a wheelchair user and wish to go to France or Belgium, Eurostar, the Channel tunnel train to Paris and Brussels, provides a good service in that you can stay sitting in your wheelchair throughout the journey. Main railway stations have a leaflet, *Rail Travel for*

Disabled Passengers, which details their assistance services and offers to help disabled travellers on and off trains.

Some companies, including Amtrak (USA), VIA Rail (Canada), SNCF (France) and NS (Netherlands), give substantial discounts to disabled passengers. A Disabled Persons Railcard will give you at least a third off all rail travel in Britain.

Road

By far the easiest way to get around is in your own car, particularly if you have mobility problems. It also gives you the greatest independence because you can stop exactly where and when you want. Discounts are available from many ferry companies. Contact the Disabled Drivers Association, the Disabled Motorists Federation and the Disabled Drivers Motor Club for information. It may be possible to rent adapted cars in some international cities, through some of the larger car hire firms such as Avis and Hertz, but check before you go.

Orange Badge schemes

A number of European countries now provide parking concessions for disabled people to include visitors holding an Orange Badge. Details of concessions available can be obtained from the Department of Transport. If there is no concession in the country you are planning to visit, take your Orange Badge and notify the local police when you get there. More often than not they will arrange parking spaces for you.

Handy hints

For holidays in the UK, invest in a mobile phone – a useful aid if you break down in a remote area. You can also phone your next destination and provide details of your route and arrival time, so that if anything happens the emergency services can find you easily.

A growing number of service stations in Europe have facilities such as specially adapted lavatories, flat access and low-level telephones. Some of them, for example France, Germany and Sweden, will supply booklets detailing these services along major roads. (See Chapter 4 for details of documents you'll need to carry.)

Sea

Ferries

Many modern car ferries offer good access, and facilities range from specially adapted cabins to lifts, good access to cafeterias and restaurants, and aids for individual disabilities such as braille menus in the restaurants. Many discounts are available for registered disabled passengers.

Cruise ships

Facilities vary. Getting on board and disembarking can be a problem, as can accessibility to facilities, such as the cinema or dining room, on board. Make sure that you check out all potential problems before you book.

Individual disabilities

Contact the organisation that deals with your disability. It will often have a range of useful leaflets and details of treatment centres and specialist doctors in the country or countries you intend to visit.

CASE STUDY *'Make the most of your support group'*

Writer and editor, Anne Lyons, 58, developed coeliac disease three years ago, which means she is allergic to gluten and, consequently, all food containing wheat, rye, oats and barley.

'My diet can be very limited when eating away from home, as not only do I have to avoid all the obvious foods such as ordinary bread, pasta, biscuits and pies but

I must also be constantly on guard against hidden sources of gluten. Even some types of curry powder contain wheat flour and will give me a severe stomach ache!

'I thought I'd probably never be able to go away again – or if I did I would not be able to eat out, as waiters obviously don't know what contains gluten and what doesn't. But my disability's organisation, the Coeliac Society, has been invaluable in providing me with explanatory leaflets translated into a range of European languages that I can send away in advance to the hotel I'm staying in. Armed with these and a small supply of gluten-free bread, I can make sure my holiday isn't ruined!'

4.
Before you go

Good research and planning are the essence of a successful trip or holiday. The more you do beforehand, the better time you will have, because there'll be less to go wrong unexpectedly and you will have a good idea of the type of place you're going to. Start by collecting a few brochures from your travel agent and browse the shelves of your local library. Read up on your proposed destination: guide books favoured by independent travellers include the Lonely Planet and the Rough Guides series. But whatever you choose to read, make sure that you get hold of the latest edition – from your local library or bookshop – as travel information and prices date very quickly while hotels and guesthouses come and go!

If you are travelling independently, work out your approximate itinerary on paper. Plan what you want to see and do – and how you will travel from A to B. Then when you come to book your journey you will have a blueprint from which to work. Of course, your plans need not be set in stone at this stage or even later; one of the joys of independent travel is the freedom it gives you to adapt your journey to suit your whims, whether it's staying on a few extra days at an idyllic spot or taking a detour to visit an extra place of interest that you hear about en route.

Getting a good deal

How much you pay for your trip or holiday depends on whether you travel during peak or low season, how far and by which method of transport. Whether you're taking an organised trip or travelling independently, transport will be one of the major expenses so it pays to shop around before you commit yourself. There are many good deals to be had when travelling by plane, especially if you take advantage of last-minute bookings. But always book through an agent who is a member of the Association of British Travel Agents (ABTA); then, if the company fails financially, the money or holiday will still be guaranteed.

The independent traveller on a limited budget can make savings by meticulous planning, first by working out which places can be reached by discounted air fares and next by picking a place where the cost of living is low. Then work out when are the cheapest times to travel to your destination – most airlines and ferry companies offer peak, shoulder and low season prices.

If you can be flexible about when you travel, you can save hundreds of pounds on long-haul flights by being a courier. There is no upper age limit, but you do have to be in good health and able to carry documents on and off the flight unaided. British Airways (see p 122) operates courier flights.

As only one courier is needed for each flight, this method of travel favours the solo traveller, although a companion can sometimes arrange to take the next flight. Most flights operate on fixed return dates of one to three weeks and cannot be changed. It's advisable to book several months in advance – although sometimes flights can be obtained at a moment's notice.

Finding accommodation

If you're taking a package holiday or organised tour, all your accommodation will be arranged for you. Otherwise, now is a good time to start deciding where you are going to stay. Most countries have a range of accommodation, from the cheapest youth hostels (not all have age limits) to guest houses, B & Bs and hotels to suit all pockets. Many women report that they feel happiest and safest in a small hotel or guest house, and indeed they're probably the friendliest option, particularly if you're travelling alone.

Whether you book a few nights to start with from the UK or decide to wait until you arrive depends on how you feel and where you're going. Popular guide books will indicate hotels and B & Bs frequented by other travellers, but, as the information dates quickly, do check out at least a couple of options before you set off. I prefer to book the first few nights of an independent trip from the UK, to avoid having to worry about it on arrival.

A note for solos

Sadly, single room supplements remain a hard commercial reality of many package holidays, although one or two operators have partially done away with them. Saga's UK holidays (see p 127) now have no single room supplements and the company is gradually introducing this policy on some of its European and long-haul tours; Solo's Holidays (see p 127) tries to avoid charging them on most of its holidays in the UK and elsewhere (with the exception of parts of the USA).

Your holiday budget

Ways of keeping the costs down – apart from booking discounted air fares and travelling off season – include opting for self-catering accommodation. If you are eating in restaurants it is often cheaper to have your main meal at mid-day, rather than in the evening when prices can be considerably higher. Alternatively, buy sandwiches for lunch and relax in a restaurant during the evening.

When you work out a typical budget for your holiday, there are two main areas to consider: transport, and expenses while on holiday.

Transport

Apart from the tickets, set aside money for transport to and from your embarkation point, any airport taxes, visitor's entrance visa fees to countries such as Turkey, refreshments while travelling, and other incidentals such as books and magazines.

Expenses on holiday

The amount of money you need to take will depend on whether you are on a package tour or travelling independently. On a package tour, you will need to budget for the meals and drinks that aren't included in the holiday price, for postcards and stamps, souvenirs and excursions, plus tips for any hotel staff, porters etc.

If you are travelling independently, add to the items mentioned above the cost of accommodation, meals, transport (local buses and taxis), based on your up-to-date guide book and advice from the country's national travel organisation.

These calculations will provide you with an estimate of how much your trip is all likely to add up to, but you should always take extra in case of an emergency. A credit card is one way of covering unexpected expenses such as extra transport or hotel bills.

In my opinion, it's also worth taking extra money, if your budget allows, to splash out from time to time on the following:

- Taxis – for the times when you're tired out from hours of sightseeing, and you just want to get back to your accommodation and relax.
- Meals – it's always fun to enjoy one or two good meals, whether you're in the UK or abroad, at a well-appointed restaurant.

Money

Cash

It's not a good idea to carry large amounts of cash with you, but do take enough of the foreign currency to pay for your transport to your hotel or destination plus a small float to tide you over in case the bureau de change at the airport or station is shut.

If you take some UK bank notes, they can be exchanged for local currency at any bank. Rates may be more favourable than buying foreign currency in the UK, and the £5 note allows you to exchange a smaller denomination than with traveller's cheques which have a minimum denomination of £10. However, the exchange rate is less favourable than for traveller's cheques.

Traveller's cheques

The most popular way of taking money abroad, the main advantage of traveller's cheques is that they can be replaced if they are stolen or you lose them. Nowadays they can be obtained in a range of foreign currencies, so you can get them in many local currencies as well as the more standard pounds sterling or US dollars. You can use local currency traveller's cheques in the same way as bank notes to settle bills, which saves queuing at the bank.

If you are travelling outside Europe, choose cheques in US dollars preferably, or failing that pounds sterling. In the Americas (North and South) dollars are essential. Take a mixture of denominations, and always shop around for the best exchange rate – this is seldom found at the airport or railway station.

Eurocheques

As the name suggests, Eurocheques and cards can be used in Europe and certain other countries as well, and are useful for drawing on your UK bank account if you need to. You can also settle bills in restaurants with them. They do, however, carry a minimum handling charge, which makes them uneconomic for small amounts and they are difficult to replace if lost or stolen. Always keep the card separate from the cheque book. There have been problems with using them in one or two European countries, so ask your bank for advice before you go.

Credit cards

These are extremely useful for emergencies and for car hire: producing a credit card can eliminate the need for a deposit in some cases. Some UK banks now offer cash dispensers in certain countries abroad. Check with your bank or the relevant national tourist office before you go to make sure that your credit cards are accepted in the countries you intend to visit.

Essential documents for foreign travel

You will need to get certain essential documents ready. Top of your list is the passport.

Passport

Most foreign countries require a full British passport: one that is valid for ten years. You should apply for or renew your passport at least one month before you travel, to allow for delays at the Passport Office (see p 125); in an emergency you can obtain one in person but you may have to wait for anything up to a day and you will have to provide documentary evidence of the date you are going to travel. Passports are also useful as proof of identification when cashing traveller's cheques or checking into hotels.

Passports that have a previous stamp from one country are not always accepted in another country if a politically sensitive situation exists between them (for example, Israel and certain Arab countries). The Passport Office can issue a supplementary passport to cover these contingencies.

When you are on your travels, it is strongly advisable to keep your passport with you in, say, a money belt, or to lock it away in a safe place. Report any loss or theft immediately to the local police and take a copy of their report to the nearest British Consular Office. It's also advisable to copy down the details or photocopy the first four pages of your passport, before you leave the UK to speed up the issue of a replacement if you lose your passport or it is stolen.

Visa

Visas may be required for stays of any duration in certain countries, such as Hungary, or for stays of more than a certain duration, for example 60 days in Portugal or 90 days in Spain. The rules can change, so you should check personally that you have the latest information for your destination. Do not rely on advice from your travel agent or tour operator, as responsibility for valid passports and visas is yours. The relevant foreign consulate or national tourist office here in the UK will oblige. A visa extension can usually be obtained abroad by visiting the relevant consulate.

Other useful documents

Driving licence

This will double as proof of identification, and you never know when you might want to hire a car. Some countries require an international driver's licence (see 'Taking a car abroad' on p 50).

Insurance

Insurance policies vary greatly in price and in what they cover. Legally you don't have to be insured when you travel, but it could be very costly if, say, you're involved in an accident or are robbed. Most people tend to buy their travel insurance from a travel agent or tour operator, but remember that banks, building societies, insurance companies and brokers also sell insurance and their rates are often lower.

Travel agents can on occasion be expensive middlemen for insurance policies. Beware in particular a compulsory insurance that is part of a discounted offer – some of these policies can cost double the cheapest rates. It may be cheaper and more appropriate to your needs to consider other options, such as a medical-only policy, if your possessions are covered by your all-risks home contents policy.

If you know in advance that you'll be taking several trips abroad in the course of a year – at least twice – many banks, building societies and specialist insurers offer an annual travel insurance policy. A free leaflet covering holiday and car insurance is available from the Association of British Insurers.

INSURANCE CHECKLIST: MAKE SURE YOU GET THE COVER YOU NEED

- First and foremost, it's a good idea to check all your insurance policies – from home insurance to credit cards – and see exactly what you are already covered for. Check that you're not insuring yourself twice over, as certain articles such as lost baggage, possessions or money may be covered by your household contents insurance.
- Many credit cards offer travel insurance, normally free with gold cards, and personal travel accident cover if you pay for the holiday with them. This pays a benefit on death, permanent total disability or loss of an eye or limb. But they will cover only the journey itself. Credit cards can also cover you against cancellation under their purchase protection insurance.
- Although there are insurance benefits on getting older, many insurers impose conditions or restrict cover for persons over 70 and premiums can be increased, so it is vital to shop around. Some companies such as Age Concern Insurance Services and SAGA Services offer specific policies for older people.
- Always check that the policy covers your needs. Read the small print. Clauses such as 'usual policy exemptions apply' may refer to pre-existing medical conditions. If you have a medical condition, you will need a special policy.

When deciding on an insurance policy, look out for the following:

- **Medical expenses** Experts recommend a minimum of £250,000 for travel to Europe; £1 million to the USA and the rest of the world. This will cover you for essential medical, dental and hospital treatment, including a flight home by air ambulance.
- **Personal liability** This covers you if you accidentally injure somebody and they then claim against you. It includes legal costs. Aim to get cover of at least £2 million for the USA – where lawsuits are expensive – and £1 million for Europe and the rest of the world.
- **Cancellation or curtailment** Look for insurance that covers you if you have to cancel or cut short your holiday if you fall ill, or if a close relative or travelling companion falls ill or dies, or if your home is severely damaged by flood, fire, storm or theft. Some policies will also cover you if you have to take a more expensive flight home in an emergency – this can be useful if you are travelling on a charter flight.
- **Belongings** Most travel policies usually cover belongings on an 'indemnity' basis, which means that, if you claim on them, a deduction is made for wear and tear. A minimum limit of £1,000 is advisable, but obviously it's sensible to work out exactly what your baggage is worth before you make any decision.
- **Expensive items** Make sure that, if you take expensive items such as a camera, you insure them specifically by naming them on the policy. If, say, the limit is £100 and you have a diamond ring that is worth £400, you should leave the ring at home.
- **Cash limits** Although the total limit for the loss or theft of money ranges from £200 to £500, the limit for actual cash is often only half this amount, so check your policy carefully. A number of policies also include the cost of replacing important documents, such as airline tickets and passports, so long as they are under the money limit.

Taking a car abroad

When driving abroad, you need to carry the following documents:

- A full UK driving licence and, for some countries, an international driving permit (obtainable from the AA or RAC).
- A green card from your motor insurance company, which provides evidence that you have insurance above the minimum. It is proof that you are insured against all damage caused to others if you are involved in a motor accident.
- A bail bond is recommended for driving in Spain. This guarantees you bail in the event of an accident; in Spain accidents can have serious consequences and cars may be impounded or drivers kept in custody. You can get a bail bond free from your motor insurers, the AA or the RAC.
- Other essential items include your vehicle registration document, GB sticker, warning triangle, first aid kit (see Chapter 6), left as well as right side-mirror, black tape for altering headlamp dip and headlight bulbs.

Car hire abroad

If you are intending to hire a car when you are abroad, the safest option is to arrange for its hire and insurance in the UK, because then you will have some means of redress if things go wrong. Make sure that your policy has Collision Damage Waiver (CDW); otherwise you will be liable for the first £250 or more of accidental damage to the hired vehicle, irrespective of whose fault it was. In the USA, third-party cover is low by UK standards, so check that your policy has 'top up' insurance.

Looking after your home

Statistics show that four out of five burglaries take place when the house or flat is empty, so it makes sense to make sure that your home is as secure as possible, even if you're only going away for a short trip. (For a longer stay away, you might like to consider letting your home, via a reputable agent, or employing a house-sitting service rather than leaving it empty.)

Before you depart for a few days or a few weeks, check your house contents insurance to make sure it's been updated to include any new purchases and also that it doesn't become invalid if the home is left unoccupied for more than two weeks, as some policies do.

Preventing burglaries

As well as making it difficult for potential burglars to break in, you can keep your home secure by fooling would-be thieves into thinking that someone is at home. Contrary to popular belief, home security needn't cost a fortune.

Security measures that don't cost a penny

- Join or set up (via your local police station) a Neighbourhood Watch scheme. Posters and stickers warning that residents are on the look-out can be a powerful deterrent.
- Ask a neighbour to drop in from time to time/mow the lawn if you have one.
- Don't tell many people in your street – or local shopkeepers – that you are going away. It's easy for the wrong person to overhear. When you cancel all papers and milk, be discreet.
- Never leave keys in locks, under mats or in the proverbial flowerpot.
- Unless you can arrange for someone to drop in regularly to draw your curtains, leave them open. Most burglaries take place during daylight hours, so drawn curtains can be a real give-away.

- Make sure that valuable articles cannot be seen from the street; put TVs and videos in a corner as much out of view as possible.
- Don't leave any labels with your home address on the outside of your suitcase. Far better to write your name and address on a piece of paper and keep it inside your case.

Security measures that cost a little

- The Post Office's Keepsafe system will keep all your mail at a local sorting office and deliver it on your return. It costs only a few pounds for two weeks. Apply at your local post office or call 0345 740740.
- Buy a time switch (available from most DIY stores) to turn the radio and lights on and off at prearranged times.
- Fit a five- or ten-lever mortice deadlock to the front door and key-operated window locks to windows that can be reached from the ground. Two-thirds of all break-ins occur at the rear of the house, so make sure that the kitchen door is well secured, too. Your local Crime Prevention Officer will provide the names of recognised locksmiths.

Precautions for valuables

- Keep valuables safe.
- Take a photograph (a Polaroid will do in most cases) of all your valuables.
- Make a note of any serial numbers on equipment.
- Fit a burglar alarm – recognised to be a good deterrent. Choose one passed by NACOSS (the National Approval Council of Security Systems).
- A free Home Office leaflet, 'Peace of Mind while You're Away', is available from your police station or PO Box 999, Station Road, Sudbury, Suffolk CO10 6SS.

When you're putting your valuables away, remember that many thieves are all too familiar with places for hiding valuables. Avoid the following:

- top of the dresser
- lingerie drawers
- top shelf in the wardrobe
- bedside table
- under the mattress
- washing machine
- freezer
- inside shoe boxes
- inside shoes
- under the lavatory seat

Source *Sunday Times* 31 July 1994

House-sitters

If you wish, you could allay all worries about security by employing a firm to provide a house-sitter, who as well as staying in your home will answer the telephone, care for pets, water the plants and mow the lawn. You can arrange for them to collect you from the airport and purchase groceries ready for your return. Companies offering this type of service include Housewatch and Homesitters while Animal Aunts will also look after a wide range of animals, including goats and horses.

Packing

Choose your clothes with care

What you take with you will depend on where you're going and what type of traveller you are, but the golden rule is to travel as light as possible. Nowadays the majority of items you may need can be purchased in most locations around the world. Useful packing tips include using underwear to fill any gaps and taking plastic carrier bags, which can be used for all manner of storage purposes.

Footwear

Most vital to the traveller is a pair of good shoes. Never go away with shoes that are not broken in – and preferably take ones that have low heels, thick soles and decent arch supports. Some of the hardier travellers swear by walking boots, but for many purposes trainers are suitable. Sandals are only appropriate for hot, sunny weather; for winter conditions you'll need boots. Remember that in certain countries the snow is salted, which will leave marks on leather boots, so don't take your smartest pair.

Basic wardrobe

Travelling as light as possible is not as easy as it sounds. Obviously which clothes you take with you will depend on the weather conditions and how you are travelling/where you are staying. Clothes should be made of natural fibres where possible – far better in the heat! If you take them in co-ordinating colours, they can be combined to form a variety of outfits. A typical basic wardrobe might include:

- 1 pair of tough crease-resistant trousers/jeans
- 1 pair of lighter cotton trousers or skirt
- 3 blouses or T-shirts
- 1 cardigan and 2 sweaters (one of sweatshirt material, one heavy cotton)
- 1 dress
- 1 jacket or raincoat
- 2 pairs of walking shoes
- 1 pair of sandals

- 1 nightgown (optional extra, one light cotton dressing gown)
- 1 swimsuit
- 1 pair of slip-on slippers (wear throughout a long flight)
- 3 changes of socks, stockings/tights
- 3 changes of underwear
- 1 scarf
- 1 umbrella
- 1 hat (wide-brimmed)

Depending on your holiday, you could buy a cheap straw hat when you're 'out there' and leave it behind; this saves worrying about whether your hat is getting crushed on the journey.

This is the minimum anyone can get by on – and together should not weigh more than a few kilos. You can easily adapt your wardrobe to the seasons and climate – adding such items as thermal long johns, warm winter coat, gloves etc for winter holidays. For a trip to a city such as London or Paris, you could replace one pair of trousers and a sweater with a smarter co-ordinating two-piece.

Other useful items

The following can be essential in a variety of situations:

- a small torch (plus battery)
- scissors
- nailbrush
- tin and bottle opener
- universal bath plug
- hair drier plus universal plug
- not your best watch
- travelling heating element, coffee/tea etc
- hot water bottle

Backpack or suitcase?

Which you take depends on what type of traveller you are. There is certainly no age restriction on carrying a backpack – although you should be reasonably fit and mobile – and it does leave your hands free. Modern backpacks have come a long way from the cumbersome iron and canvas constructions that were popular 30 or more years ago, and, although they are not cheap, some can now be purchased that double as a suitcase. Collapsible nylon suitcases with a zip fastening are also worth considering for shorter breaks – they weigh next to nothing.

The size of your luggage will depend on how you are travelling but try to keep it as small as possible. A big suitcase can be impossible to move – unless you invest in a pair of sturdy wheels (a pair of flimsy ones is worse than useless) – and may not fit on to luggage racks in trains and the like.

Always take one piece of hand luggage that contains everything you need for the night, plus any medication, guidebooks etc, in case your main luggage gets lost.

If you are travelling independently by train, make good use of luggage lockers. When you arrive at a station, use them to store your belongings while you look for somewhere to stay.

5.
Staying safe

Any travel involves a certain amount of risk but it is important to keep potential problems in perspective and not let needless worries overwhelm you. Although isolated incidents receive a great deal of media coverage, do not let them put you off travelling in the UK or abroad. With proper precautions, the risks are minimal – as the thousands of people who return unscathed each year can testify.

Obviously it is sensible to avoid travelling to countries where there is prolonged civil war or serious medical problems such as an outbreak of cholera or malaria. It goes without saying that walking or hiking alone in isolated areas is highly dangerous and should not even be considered at any age, nor should you ever accept invitations from strangers. Furthermore, never agree to carry anything for a stranger or recent acquaintance.

Common sense should guide all your actions. For example, it is better to arrive at a destination during daylight hours – particularly if you have not arranged any accommodation – rather than late at night. When you arrive at a new destination, always ask about local conditions: 'Where is it safe to go?'; 'Are there any problem areas where women are advised not to venture?'; and so on. Staff in hotels, guest houses and hostels, as well as tour company representatives, know all about local conditions and are usually extremely helpful.

Sensible precautions

Other basic ways to make sure you stay safe – which can be applied to most places, whether you're travelling alone, with a companion or in a group – include the following.

Blend in

Try to blend in with your surroundings as much as possible. The obvious tourist trappings of guidebooks, cameras, luggage and, often, choice of clothing mark you out immediately as a possible target for touts and pesterers.

As a general rule, the farther south and east you travel, the more conservative the dress for women. This is not so much the case in tourist areas, but, if you venture off the beaten track, a sun dress might provoke disdain or even hostility.

Although most places in the world are safe to visit, there are exceptions. Multicultural Muslim countries such as Indonesia and Malaysia are safe to visit alone, but middle eastern Muslim countries have to be treated with care, especially by single women travellers. If you wish to see these countries, it's advisable to join a tour or organised group.

A good way to disguise the inevitable camera and guidebooks is to keep them safe in a large shoulder bag, which should have a flap and a secure fastener. Some travellers recommend cutting out the relevant sections from guide books; a neater solution is to photocopy the relevant sections from your book before you go. Always carry your bag on the side away from the road.

Know where you are staying

Always know where you are staying. Keep a map of the area in your pocket, and the address and telephone number of the place you are staying at, along with the taxi fare back there. Take the card and telephone number of a reputable local taxi company with you, if you wish.

Be confident

Timid people send out signals that, psychologists say, encourage attacks. If you're walking anywhere alone, walk purposefully, with your head up and shoulders back. Always try to look as if you know where you are going. Get into the habit of planning your route beforehand. Carry the relevant pieces of the map, which will save you from having to stop and ask for directions. Avoid looking like a tourist by never standing with a map flapping in the wind on street corners – if nothing else, you may attract touts and beggars.

Never keep all your money in one purse

Don't carry large amounts of cash around; it's better to change traveller's cheques as necessary if you are abroad, charge tickets and other purchases to a credit card, or use a cheque book when in the UK. Keep cash separate from credit cards, traveller's cheques, airline tickets and your passport. You might consider investing in a money belt or, at worst, you could tuck some of it away into your bra. Be on your guard, especially in crowded places such as on public transport or in busy streets, and don't hang your bag on the back of your chair in restaurants.

Have a contact name and number

Wherever you go, it's a good idea to carry the name and address of someone who can be contacted in an emergency. Whenever you arrive at a destination, find out the emergency numbers for ambulance, fire and police – you probably won't need them, but it's best to be prepared.

Luggage

If possible, don't take shiny new luggage – if only because it makes your property stand out as a potential attraction. Neither should you attach an address label to the outside of any suitcase or bag; keep it safely tucked inside. A combination lock is often safer than a standard one that may be unlocked by many a key. Keep your luggage next to you at all times: if you go to, say, the dining car on a train, take your money, passport and tickets with you.

Telephone

In a public telephone box don't rest your handbag between your legs or leave it on the ledge; keep your wrist or shoulder through the strap and stand facing the street. This will prevent it being snatched.

Accommodation

Hotels and guest houses are generally safe places, but it's common sense to take the following precautions while staying in these types of accommodation.

- Don't leave valuables lying around in your room. Take a travel lock as an extra precaution and lock valuables in your case before you go out, or use a hotel safe if available.
- If there's a knock at the door, always ask who it is. Do not open the door to anyone until you are sure of their identity.
- Check that windows and doors, particularly those that back on to a balcony linking other rooms, are securely fastened when you go out and at night. It's far better to spend the night in a stuffy atmosphere than run the risk of being robbed or worse.

Tips for travelling alone

In a car

When driving alone, observe the following guidelines:

- Let people at your destination (eg friends, or the hotel or guest house) know your approximate arrival time and your proposed route.
- Avoid lonely areas. Make stops only at busy laybys and petrol stations. Always lock doors even if you'll be away from the car for only a minute or two.
- In isolated areas drive with all the doors locked; otherwise drive with the passenger door locked but your door unlocked, in case of an accident. Always keep your windows high enough to prevent someone reaching inside.
- If you have to stop and wait at night for any reason, move over and sit in the passenger seat, advises Elizabeth Green, because 'It looks as though someone will be joining you.' She also keeps a man's hat in the car, to maintain the same illusion.
- If you have to leave your car to get help, do not accept lifts from a man on his own or with other men. Use your common sense in deciding whether to accept a lift from a woman, or a man and woman together.
- Do not leave your handbag lying on the front or back seat. Tuck it away out of sight.
- Carry a personal alarm as an extra precaution when you enter and leave the car after dark.
- Do not stop if you see a car that appears to have broken down or had an accident, unless you have witnessed it happening: it could be a trap. Drive on to the next telephone or police station and report it.
- If someone tries to flag you down by indicating that something is wrong with your car, do not stop but keep going to the next service station or police station.
- Don't pick up hitchhikers if you are alone.

- A car telephone enables you to call for help from the safety of your vehicle, if you break down, and to notify people if you are delayed or in trouble of any sort.
- The AA produces a useful card, 'Guidelines for women driving alone', which you can keep in the glove compartment. Available free of charge from the AA on 01256 492195.

If you are unfortunate enough to have an accident:
- Don't move the car until the police arrive.
- Put out the warning triangle.
- Help anyone who is injured but do not move someone who might have a serious fracture or back injury – you could make it worse.
- If you have a camera, take photographs from several angles.
- Do not sign anything – if the police insist that you do, write by your signature that you do not understand what the form means.
- Get a copy of the police report; you will need it for insurance purposes.

On a train

Travelling in a crowded carriage, while not necessarily the most comfortable way to see a country, does usually guarantee your safety. However, you should:
- Avoid getting into an empty carriage – you never know who might get in at the next stop. If there is no alternative, pick a carriage that is next to the guard's van, usually at the rear of the train.
- Avoid travelling in a carriage where there is only one other person present.
- Avoid railway stations at night. Some of the larger ones in European cities, for example, attract a rather odd mixture of people in the evenings and at night time, although they are perfectly safe during the day.

On foot

There's safety in numbers, as very few people are attacked if they are in a group or even accompanied by one other. Take advice from your hotel, tour company representative or up-to-date guide book. Again use your common sense and take note of the following, if you are walking alone.

- Don't take shortcuts along unlit or deserted roads.
- Stay away from lonely alleys, back streets and the like.
- Face oncoming traffic to reduce the risk of kerb-crawlers.
- If you think you are being followed, cross the road and walk on the other side. Should the person(s) follow you, head for a busy place or police station. If you are in a residential area, walk towards the nearest well-lit house; most followers will not wait around to see if you are let in.
- Street sellers and stall holders in some countries try to sell their wares by pestering as many people as possible. Stand your ground and repeatedly but calmly say you are not interested. They will eventually give up. I also find that not catching the eye of the salesperson is another useful method of evasion.

If you are robbed

If you are so unlucky as to have your purse or bag stolen, report the crime to the police at once. Get a copy of the report they file (you will need this for insurance claims).

Credit card loss is now becoming a problem to be reckoned with. According to one card loss-reporting service, the worst five plastic blackspots in Europe are Italy, Spain, the Netherlands, France and Germany, and cities to watch out for world-wide include Prague, Paris, Budapest, Alicante and Hong Kong. Always keep your cards separately from your money and traveller's cheques. Reporting stolen cards as soon as possible may mean, under the Banking Code of Practice, that you are liable for only the first £50 of fraudulent use. But it's better to be safe than sorry, so:

- Take care when using cash dispensers – some are 'staked out' by muggers.
- Remember that you're more likely to have your cards stolen at the beach or by the pool, in bars and restaurants, in your hotel and while travelling – especially at airports.
- Consider taking out a card protection policy, which means that you report the loss of cards to one location rather than having to deal with them all individually.

6.
Staying well

The main areas of health that any traveller to foreign parts should be concerned with are immunisations, food, drink and hygiene. Thanks to the advances in modern medicine and equipment, people with physical disabilities (see Chapter 3) can travel more freely than ever before; similarly, people with chronic illnesses such as heart disease, diabetes or a respiratory condition need not think that they are bound to stay home.

Before you leave

However well you may feel, if you're over 60 or suffer from a particular complaint, it's a good idea to discuss your proposed travel itinerary thoroughly with your doctor. Make sure that you take enough medication with you – in your hand luggage, in case your suitcase or backpack is lost.

It's also sensible to take along a short medical history and details of your prescriptions. Your doctor will write down for you the names of your medicines – their brand names and generic (chemical formula) names – and what each one is for. People are often unaware of what their tablets do, and, if they are taking more than one, which tablet does what – now is a good time to find out! Whilst the brand names may differ, the generic names are usually the same or very similar throughout the world (although in the USA, paracetamol is called acetaminophen).

Don't worry too much about the language your prescription is written in. In many developing countries, the pharmacist is highly educated and will almost certainly speak English or French and will be able to interpret an English prescription. If you wish, however, you can get your prescription translated in the language(s) of your destination country(ies).

If you wear spectacles, take a spare pair and get a copy of the prescription from your optician. It's also a good idea to take a pair of prescription sunglasses – with polarised lenses. If you use contact lenses you should take plenty of cleaning and soaking solutions. Most major brand name lens cleaning solutions are available worldwide (except in remote areas) but do be prepared for possible eye irritation in a warmer, dryer or dusty climate, or in the dry, air-conditioned atmosphere of an airplane cabin. Some people who normally use contact lenses prefer to travel wearing their specs – that way, they can take the occasional nap on a plane or train without having to worry about lenses and bottles of solutions.

Special precautions to take

- **Heart disease**: avoid high altitudes and strenuous activities. Pacemakers can trigger the alarms at airports, so inform the security guards if you have one.
- **Diabetes**: plan your insulin schedule with your doctor, taking into consideration any time zones you must cross. You may need to test your control more frequently than at home, even if you only take tablets or control your diabetes by diet.
- **Asthma** and **chronic bronchitis**: these are examples of respiratory problems that are often made worse by changes in temperature, humidity and altitude. Check with your doctor that your itinerary is suitable for you. With some lung conditions, travelling in an aircraft can cause problems.

Prepare early

At least two months before you go, see your GP about any immunisations you will need and arrange to begin courses of any medication (such as anti-malaria tablets) at the right time. The booklet T5, *Health Advice for Travellers*, published by the Department of Health is available free by phone (Tel: 0800 555777) or from main post offices and some travel agents. British Airways Travel Clinics (for details of your nearest clinic call Talking Pages free on 0800 600900) or MASTA, part of the London School of Hygiene and Tropical Medicine (0171-631 4408), can also provide advice. MASTA also have a 24-hour Travellers' Health Line, which offers written details of immunisations and malaria relevant to your journey (Tel: 0891 224100; calls cost 39p per minute cheap rate, 49p per minute all other times).

If you intend to visit an EU country, get form E111 – the application is in booklet T5 (see above) – which entitles you to the same state health care as a national of that country. Sometimes this means that there may be a charge, part of which might be refundable later. But you will still need travel insurance (see Chapter 4) to provide repatriation, should that be necessary.

There are various reciprocal health agreements with some countries outside the EU, including New Zealand and Australia, which will give you some cover. Again, leaflet T5 has all the information. If necessary, ask your doctor for a full check-up before you go; if you intend being away for more than a few weeks, also include a visit to the dentist. (Many insurance policies cover only emergency work on teeth, and you may save yourself a great deal of pain and expense.) Similarly, a trip to a chiropodist will ensure that your feet are in good condition, ready for any additional walking that your trip may entail.

Immunisations

These can be carried out by your local GP; British Airways Travel Clinics also offer this facility, but they charge. The Department of Health leaflet T5 gives a detailed country-by-country checklist of the immunisations – they can be divided into compulsory and recommended.

Compulsory immunisations

'Compulsory' here simply means that the country concerned will not allow you to enter unless you have a valid certificate of immunisation, signed and sealed by the doctor. Yellow fever and cholera fall within this category.

Yellow fever

Caught from the bite of an infected mosquito, this serious disease (it attacks the liver and causes jaundice and fever) is found in Central Africa and South America. It is essential to have a vaccination certificate if you are going to these parts, and also if you are travelling on to certain other countries, such as Australia, afterwards. The vaccine can only be given at a Yellow Fever Vaccination centre – see your doctor for details – and is not available on the NHS. It becomes effective after ten days and lasts for ten years.

Cholera

This is a bowel infection causing severe diarrhoea that in turn can lead to dehydration and death. Found in many countries in South America, the Middle East, Africa and Asia, it is caught from the consumption of contaminated food and water. A vaccine is available but it gives very little protection: the best way to avoid cholera is by scrupulous attention to food and personal hygiene. Although officially no country now asks for a compulsory certificate of vaccination, travellers to cholera areas who may cross remote borders overland are recommended to have it, to minimise the possibility of border problems. The injection lasts for six months.

Recommended immunisations

Recommended immunisations are for hepatitis A, polio, tetanus and typhoid.

Hepatitis A

This is a virus infection of the liver, causing jaundice, that is usually caught through consuming contaminated food or water. A new vaccine, Havirex, is now available but it's expensive; and the older immunoglobulin (gammaglobulin) immunisation is now used only for people going on a short, one-off trip. You need two injections before you go, and one six to twelve months later will give ten years' immunity.

Poliomyelitis

Caused by a virus, polio may result in paralysis. If you intend to travel outside north and western Europe, North America, Australia and New Zealand, you should make sure you're immunised against it. For most of us this will mean a booster dose, taken as drops on a sugar lump. If for some reason you have never been immunised, you'll need a full course of three doses of vaccine.

Typhoid fever

Like cholera, typhoid is caught by consuming contaminated food and water. Its symptoms include headache, fever, constipation or diarrhoea and a rash on the chest or abdomen. Immunisation against it is recommended if you are visiting countries where sanitation is poor; it is available in both injection and tablet form.

Tetanus

Caught through the introduction of bacterial spores through a wound (even a tiny cut), tetanus causes extremely painful muscle spasms (hence its nickname 'lockjaw'). Tetanus is particularly dangerous in remote areas where medical treatment may not be immediately available. Most people will have been immunised in childhood, but a booster is needed every ten years. If you have never been immunised, you will need a course of three injections.

Other precautions
Malaria

Spread by the bites of infected mosquitoes, this parasitic disease causes fever, and sometimes complications affecting the kidneys, liver, brain and blood. Endemic in parts of the world, notably the tropics, it can be fatal. There is no vaccine for malaria. Instead, travellers should take prophylactic drugs (that prevent the disease from starting), usually beginning one week before they go and continuing for a month after they return – or leave the area – to get rid of any parasites that are still present. As some strains of malaria are resistant to anti-malaria medicines, you should tell your doctor which countries you plan to visit, to make sure you get up-to-date advice. The London School of Hygiene and Tropical Medicine operates a Malaria Helpline (0891 224100), providing information on which malaria prophylactics are needed for your destination. (A call costs 39p per minute cheap rate or 49p per minute at all other times.)

In areas where malaria is present, you should make sure that you keep your arms and legs covered at dusk (the time mosquitoes are most active) and use an insect repellent – preferably one containing DEET (diethyltoluamide) – on any exposed skin. Insect repellent coils are sold all around the world in markets and local shops: they are the cheapest and easiest way of keeping mosquitoes at bay while you are sleeping. You could invest in a mosquito net, but these are heavy and bulky.

Common travellers' complaints

Accidents

Many accidents are avoidable but, nevertheless, more travellers die from them than from any other cause. Flying is one of the safest forms of transport, although you should make sure that you pay attention to the safety talk at the beginning of flights. When driving, always wear a seat belt and do not drive under the influence of alcohol or when over-tired. When staying in hotels or guest houses, always make yourself aware of emergency exits, fire extinguishers and other safety features. Hotel and apartment balconies can be unsafe, so do not step out on to them if you have the slightest doubt! Finally, do not swim after you have had alcoholic drink.

AIDS

This ultimately fatal disease continues to increase world-wide. AIDS (Acquired Immuno-Deficiency Syndrome) is caused by the Human Immunodeficiency Virus (HIV). AIDS leaves the human body's defence system vulnerable to a wide variety of diseases that would not otherwise affect it. It is transmitted mainly through sexual contact or contaminated blood that comes into contact with your blood. Protected sex (using a condom) is essential; be careful with injections abroad – take your own disposable needles and syringes when going to countries where health care might not be of the same standard as in the West (see 'Basic medical kit' on p 76); there is also danger of infection from blood transfusions in some countries. Avoid skin-piercing procedures such as acupuncture or ear piercing.

Athlete's foot and other fungal skin infections

Fungal skin infections thrive in warm, moist conditions and are found most commonly between the toes, in the armpit or the groin. Itchy athlete's foot is the most common infection, spreading between the toes and on the foot. Help prevent it by wearing flip flops when in the shower and frequently wash and dry your feet carefully. Change underwear (which should preferably be cotton) and socks daily. A medicated talc can help prevent the condition, or clotrimazole (Canesten) cream can be used to treat it.

Thrush is a fungal infection of the vagina, which may cause intense itching and discomfort. Caused by temporary over-production of the body's yeasts, it is often accompanied by a white, cottage-cheese-like discharge. If you're prone to thrush, try to prevent its onset by wearing cotton underwear and loose-fitting trousers or a skirt. Always wash between your legs front to back to stop yeasts moving from the bowel into the vulva. Clotrimazole (Canesten) cream or Nystatin pessaries should clear up an infection. You may also care to consider a natural method of prevention: plain yoghurt contains bacteria which fight the yeasts – eat plenty of it while cutting down on carbohydrates and alcohol. Plain yoghurt can also be applied to the affected area to soothe the itching.

Bites and stings

Insect bites are one of the commonest complaints affecting travellers. Mosquitoes and mosquito-like insects tend to gather near water and in shady woods. Preventative measures (see 'Malaria', p 70) and mosquito coils are musts if you intend visiting these areas. To soothe insect bites, apply an antihistamine cream or even calamine lotion. Avoid scratching them, to prevent infection.

Stings from bees or wasps are not so painful, unless you develop an allergy. Gently scrape out an embedded sting and apply antiseptic cream. People who are allergic should carry an identifying Medic Alert tag, the form for which can be obtained from your GP.

Animal bites from dogs are the most common, although when you are travelling in tropical or safari areas take safety precautions and avoid contact with all mammals. All bites carry a heavy risk of infection, so, if you are bitten, the affected area should be thoroughly cleaned at once with soap and running water, and immediate medical treatment should be sought to take precautions against infection, rabies and tetanus. Apply iodine if you have it with you.

Despite the tales that abound, few snakes are poisonous. If you are bitten, do not cut or suck out the bite, but keep the area of the bite still, stay calm and seek immediate medical help.

Constipation

Changes of diet and a lack of fluids can cause constipation, as can too little exercise. Drinking plenty of liquids and eating fresh fruit and vegetables is better than taking laxatives.

Cuts and grazes

Cuts and grazes should be cleaned with clean water and bathed with iodine before being dressed. In the tropics, bacteria multiply much faster than in a temperate climate and even small grazes can fast become big problems – so treat all cuts as soon as possible, however small.

Cystitis

Cystitis is usually caused by bacteria from the anal region contaminating the urinary passage. Symptoms include frequent and painful urination, and the best treatment is to drink plenty of fluids, including orange (or other citrus fruit) juice, as soon as an attack starts. This acidifies the urine and stops the bacteria multiplying. If the symptoms don't go away, see a doctor because you may need a course of antibiotics.

Diarrhoea

Most diarrhoea is caused by food or drink. In most cases, recovery will take from 48 to 72 hours, and you must make sure that you drink plenty of non-alcoholic fluids to prevent dehydration. If your diarrhoea is severe, you should aim to try to replace the salts your body is losing, either by a proprietary oral rehydration sachet such as Dioralyte or by making your own solution by mixing half a level teaspoon of salt with eight teaspoons of sugar in a litre of water. Over-the-counter remedies such as Imodium may also be helpful. MASTA recommends a 500mg dose of ciprofloxacin, available on doctor's prescription only (and not suitable for people with some conditions, such as epilepsy) to help reduce the duration of traveller's tummy, in some cases clearing it up completely. (See sections on 'Safe eating and drinking' and 'Hygiene' on p 77 for ways to avoid diarrhoea.) If the diarrhoea continues for more than five days, seek medical advice.

Heat exhaustion

Heat exhaustion occurs when you do not drink enough liquids or consume enough salt to replace what you lose through sweat. Symptoms include nausea, headaches, tiredness and light-headedness. Rest and drink plenty of fluids, such as fruit juice or slightly salted water.

Jet lag

Jet lag is a feeling of fatigue and disorientation, often experienced by air travellers when they cross several time zones (from east to west or vice versa) in a few hours. Unfortunately, the older you are the more you tend to suffer, so, if you can, take a few days at the end of the journey to acclimatise yourself to the new time. The body adapts to time changes at roughly one hour per day, so if you have crossed six time zones it will take six days to adjust fully. Help prevent jet lag by avoiding alcohol, tea and coffee when flying, and drinking as much water as possible. If you can take daytime flights, these cause least fatigue and loss of sleep: when you arrive, reset your watch, and eat meals and go to bed at the new time. If you have problems adjusting, your doctor may be able to prescribe a mild sleeping pill.

Sunburn

Sunburn is a common complaint that is best prevented by covering the skin at the first hint of discomfort, and wearing a sun cream with a high enough sun-protection factor. (Note, however, that it protects only against sunburn, not ageing or skin cancer.) Cold water and calamine lotion will help soothe sunburned skin.

Travel sickness

The symptoms of travel (or motion) sickness include nausea and vomiting as well as cold sweating and is most commonly suffered in cars and/or boats/ships. There are a number of over-the-counter remedies available, ranging from antihistamine tablets to wrist bands, but none completely prevents symptoms in everyone. If at all possible, lie down or fix your eyes on the horizon or other fixed point outside.

Essential medical supplies

Although stomach upsets and too much sun remain the chief ailments suffered by travellers, other common hazards include falls, animal bites and stings, and road accidents. But should you buy a medical kit or make one up yourself? Again, money, time and method of travel as well as destination will influence your choice. On the whole, though, bought medical kits can be very expensive and they do vary in quality and range of equipment. If you're going to buy a kit, decide what you want – a light kit to keep in a suitcase or backpack or a heavier one that copes with all emergencies to keep in a car boot. If possible, check the contents and look for clear, unambiguous diagrams and instructions before you buy.

Alternatively, the following will serve as a do-it-yourself kit.

Basic medical kit

A basic medical kit should include:

- pain killers (aspirin or paracetamol)
- antiseptic cream
- antihistamine cream for bites and stings
- lip balm
- travel sickness pills
- sun protection cream
- a bottle of calamine lotion
- adhesive bandage strips

A plastic kitchen container with tightly fitting lid is ideal for storing it all in.

Possible additions

Add as many of the following as are relevant for trips to warmer regions and/or third world countries: mosquito repellent; mosquito net; malaria pills; water-sterilising tablets*; anti-fungal medication; oral rehydration sachets; antibiotics for diarrhoea; thermometer; a set of sterile needles and syringes to be used by local medical personnel in an emergency (eg for blood transfusions).

If you're unsure about putting basic first aid into practice – and particularly if you're going somewhere off the beaten track – you might consider taking a short course. Both the British Red Cross (Tel: 0171-235 5454) and St John Ambulance (Tel: 0171-235 5231) can give advice on cheap and widely available courses in the UK.

---

* If you wish, a bottle of tincture of iodine (also called 2 per cent or first aid iodine) can be used to sterilise water, but do not use for more than one year. Add three drops to one litre of water and leave it for half an hour before using

Safe eating and drinking

While there is no reason why you should not sample the local dishes and delicacies of the places you visit, do exercise proper care and caution when choosing food, particularly when you're in a hot climate. The basic principle is to eat only food that has been freshly prepared and cooked. Avoid tepid food and eating anything with a high risk of contamination, such as shellfish from the heavily polluted Mediterranean and Adriatic coasts, which carry more risk than, say, shellfish from the Atlantic coast. In hot regions or places where the conditions are unhygienic, don't eat uncooked foods or salads. Peel all fruit before eating it.

Contamination in water can be passed on, not only through drinking water but also in washed salads, ice cubes and ice cream. In hot regions where conditions are unhygienic, assume that all water and ice are unsafe unless you are informed to the contrary. Buy bottled water or purify your own with water-sterilising tablets. In an emergency, boiling water for five minutes at sea level should purify it.

Hygiene

Practising scrupulous personal hygiene when you are travelling – washing hands before meals and after going to the loo – will help keep you well, as will washing yourself and your clothes regularly. Wet wipes can be carried for the times when soap and water are unobtainable. In areas where the water is unsafe, clean your teeth and wash toothbrushes in bottled or purified water.

If you're ill

In a reasonably mainstream destination, on a package tour for instance, the tour company rep should be contacted for help with obtaining medical attention, coping with bureaucracy and any insurance problems, and to liaise with your insurance policy's specialist assistance company.

Insurance companies mostly employ specialist assistance services in connection with their travel insurance policies to liaise between the patient, their family and the medical team. Contact the emergency operations number: if necessary, an agent will take over and deal with the doctor or hospital and arrange payment of medical bills.

Some credit card services such as Barclaycard offer an international rescue service for card holders. American Express has a global helpline with an English-speaking doctor on hand and a network of recommended practitioners locally. The advice given is free and any resulting medical costs can usually be claimed against insurance.

The British Embassy has a list of local doctors but it will not pay medical bills. If you are in a country outside the European Union and with no British Consular presence, you can seek help from the embassies or consulates of other EU member states.

If you're worried about having to consult a local doctor – perhaps because of the language problem – a way round this is to become a member of the International Association for Medical Assistance for Travellers (IAMAT), a charitable organisation which offers a list of English-speaking physicians in more than 140 countries worldwide. Intermedic also supplies a list of recommended English-speaking doctors, available to members (subscription US$6).

On the whole, health care in many Western countries is of a very high standard, but if you're travelling outside western Europe, New Zealand, Australia, North America and Japan you may find a lack of modern equipment and drug supplies. Consult your GP before you go, particularly if you have any recurring conditions, such as cystitis or bronchitis.

Teeth

If – despite your pre-journey visit to the dentist – a filling or crown falls out, it's far better to contact a local dentist than try to deal with it yourself. For a temporary measure you can purchase a kit for a few pounds in a UK chemist's, which will stick a broken crown back in place until you can find a dentist.

7.
Holiday possibilities

Nowadays it's almost as difficult to choose the type of holiday you want as it is to decide where to go. On package holidays you can stay in one place or more than one in multi-centre breaks, you can travel by ship, rail, coach, car or plane, or combine the last two in a fly–drive holiday – the possibilities of combining options are seemingly endless. Apart from Saga, the major tour operators offering holidays for the older person include Thomson, Cosmos, Horizon and Falcon, and, of course, there are thousands of other general packages and tours for people of all ages.

But if you seek more than a couple of weeks in the sun, here are some suggestions for interesting breaks within the UK, Europe and the rest of the world. The holidays mentioned are all suitable for people over 50, but sometimes people over 65 may be required to provide medical evidence of fitness.

The retreat

Take a break from the demands of the world and get away from it all to think things through at a retreat. A time of reflection, rest and renewal is offered at a variety of centres throughout the UK, not all of them religious establishments. You do not have to be religious to go on a retreat – although places of retreat tend to be Christian, Buddhist or New Age in persuasion. You can go either alone as an individual or to a group, which is often based around a spiritual theme or a form of meditation. Some have been specially designed to help you unwind. They often provide particular comfort to those who have undergone some personal trauma, but many people enjoy them so much that they return again and again. Many Christian and Buddhist retreats refuse to put a price on the stay, asking only for a donation; if they do charge, the fee is modest. If you are over 60, there is usually a reduction. Expect to pay a commercial rate in New Age centres, however.

Information about retreats can be obtained from the National Retreat Association, the Buddhist Society, and the National Council of Hindu Temples. Cruse – Bereavement Care also carries details of retreats in its 'Holiday Ideas' leaflet.

Low-cost breaks

It is possible to see another part of the country for very little cost, if you are willing to volunteer your services. For example, the Aylesbury firm Homesitters employs people between the age of 40 and 68, who are not in full-time employment, to house-sit in a variety of locations around Great Britain – from old mill houses or country cottages to city houses and flats. All food and travel expenses are paid, as is a modest weekly payment. The company gives preference to non-smokers who have their own transport.

Similarly, a charity for the disabled, the Winged Fellowship Trust, needs volunteers (upper age limit 75) to spend a week or two at one of its five holiday centres for people with a severe physical disability. You'll need to be fit, but otherwise no experience is necessary. The Trust provides free bed and board and pays travel expenses within the UK. One recently bereaved volunteer wrote, 'It opened up a whole new life for me'.

If you're interested in environmental matters, help conserve the countryside on a weekend or week-long break with the British Trust for Conservation Volunteers (BTCV). Volunteers help with clearing woodlands and learning age-old country skills such as dry stone walling and footpath repair. More than 500 'Natural Breaks' are available all year round throughout Great Britain in areas of outstanding natural beauty. You pay a modest sum in return for all accommodation and food. The BTCV is open to people of all ages and, although some of the accommodation can be quite basic, breaks that have 'Superior Accommodation' will guarantee you a comfortable bed and shower. Book out of school and student holidays if you want to be with people of your own age.

Rent a property

There are hundreds of country cottages and houses for rent throughout the UK, in beauty spots ranging from Land's End to the Highlands of Scotland. Many of them are let out cheaply during the low season – you can find bargains during months such as April and September, for instance. Contact your local Tourist Information Centre (see your local telephone book) for details of regional tourist boards, who can assist with a list of properties. Other companies include Blakes and English Country Cottages.

If you have a yen to stay in a historic residence, The Landmark Trust, which restores buildings of architectural interest, has an amazing range of properties to let. Choose from forts, chapels and follies – there's even a pineapple-shaped house near Stirling in Scotland. All are in good condition and have heating – although obviously the efficiency of this will depend on the age and type of property. It's a good idea to enquire first about any difficulty of access. Discounts may be offered in the winter season.

The National Trust also has a range of holiday cottages in England, Wales and Northern Ireland. Generally these are cottages belonging to former estates and farm buildings owned by the Trust, although on its books are a castle folly and a former water tower. Its short off-peak breaks (from October to March) are especially good value.

Pamper yourself

Single people of all ages, as well as their friends, are always welcome at health farms, which nowadays are far removed from their carrot juice and physical jerks image. Although they do offer assistance with weight loss, the emphasis is now on relaxation and enjoyment rather than strict diets, and most offer a range of treatments ranging from relaxing massage, steam and sauna treatments to gentle exercise and relaxation techniques. Many of them also provide beauty treatments and alternative therapies, such as reflexology. Although they can be costly, many health farms have special offers – in the off-peak season and mid-week breaks, for instance. *Healthy Breaks in Britain and Ireland* describes more than 180 health farms, hotels and self-catering apartments with facilities for treatments.

Study/special-interest holidays

It's possible to find a break – from a few days or weekend to several weeks – to suit almost all conceivable interests at any time of year. They range from astrology to yachting, taking in such diversions as bowls, dancing (from folk to ballroom), rambling and walking, card games such as bridge and whist, scrabble, and more unusual pastimes such as making cane seating or murder mystery weekends where you have to work out 'who done it'. Some tours concentrate on history or wildlife, others on art and literature. Many are located in the UK, others in Europe or occasionally further afield. Major operators include Acorn Activities (for UK holidays only), Saga Holidays and Solo's Holidays, as well as the Association of Cultural Exchange Study Tours and the University of the Third Age (U3A).

Activity and sporting holidays

So long as you are fairly fit, there is no reason why you shouldn't take part in all types of physical activities. If you're feeling particularly adventurous, the YMCA National Centre in the Lake District runs a range of adventure holidays, one of them specially for people over 50. Activities on offer include canoeing, orienteering, archery and rock climbing for beginners. One lady in her 60s returns year after year for a course, according to the organisers.

If you feel like stretching yourself to the limit, consider an Outward Bound course for the over-50s. Courses include an eight-day programme of rock climbing, fell walking, abseiling and orienteering; an expedition course consisting of eight days of cutter sailing among the islands and lochs of the beautiful west coast of Scotland; and an eight-day Highland Explorer course walking 250 miles in the mountains.

The Youth Hostel Association – which has 250 hostels countrywide, many of them offering a comfortable standard of accommodation these days while still remaining modestly priced – welcomes older people and offers special-interest breaks based around activities such as bird-watching and yoga.

So long as you have no health worries, rambling through the countryside in Britain or in other countries is a popular way to stay fit. As these holidays often attract a large proportion of singles, they can be a good place to make new friends. Of course, it's always possible to form your own group and arrange your own trip; if, however, you'd prefer an organised tour, many travel companies run them, including Ramblers Holidays and HF Holidays.

Skiing continues to grow in popularity as a holiday option for the over-50s. Provided you are in good health, there is no reason why you should not continue to enjoy an annual trip to the ski slopes – or even take it up for the first time. Britain has far more dry-ski slopes than any other country, and these make a good place to start. Or you might like to join a small party of beginners going to Scotland or abroad to Austria, Switzerland, France or Italy. If you have no one to go with, Travel Companions can put you in touch with other skiers. The Ski Club of Great Britain runs a number of holidays especially for the over-50s. It also produces a booklet of exercises called *Ski*

Legs: Pre-ski exercises and runs over-50s clinics at a number of dry-ski slopes to help people get their ski legs prior to their holiday.

You might like to consider taking up cross-country skiing, also known as Nordic skiing. It's a gentler pursuit than the downhill variety because, as you ski up hill and down dale at a steady pace, you run far less risk of colliding with someone or careering out of control down a slope. But you do need fairly strong legs! Many downhill resorts have cross-country trails, and, of course, cross-country remains one of the main winter sports in Scandinavian mainland countries. Waymark Holidays, one of the few dedicated cross-country operators, can provide information.

Living with a family

Experience all the joys of family life as lived by the French with En Famille Overseas, which will organise stays of a week or two, sometimes longer, with families all over France. Places where you can stay as a paying guest, with or without a friend, range from Annecy, in the beautiful Haut Savois region or the picturesque port of Vannes on the west coast, as well as cities such as Paris. You will have to make your own transport arrangements.

For other holidays involving home-stays with other nationalities, the Central Bureau for Educational Visits and Exchanges produces a guide called *Home from Home* (£8.99). For home-swaps, see 'A home away from home' on page 86.

Camping and caravanning

Whether you want to travel in the UK or on the Continent, this type of holiday offers great flexibility. Many sites offer a high standard of on-site toilets, showers and laundry facilities. Choose the time of year to go with care – you will find that sites in both the UK and the

Continent are not too crowded during May, June and September, and discounts may sometimes be obtained.

If you don't own a caravan or tent, one solution is to try a package holiday with a company such as Eurocamp, which has 240 sites spread throughout Europe, ranging from Sweden in the north to the island of Corsica in the south. You can either drive there or relax and take the car by Motorail to a station near your site. Stay in a comfortable tent sleeping between two and six with all mod cons or a mobile home that can sleep between six and eight.

Many sites are situated in areas of great natural beauty. Places to consider include the Black Forest in Germany, Tuscany in northern Italy and parts of northern Spain.

The National Caravan Council can provide free brochures detailing the sites of six regions in the UK where you can rent a mobile home.

A home away from home

One way of avoiding hotel bills or rent for self-catering apartments is to exchange your home for someone else's, reducing the cost of your holiday considerably. You can always make private arrangements through clubs and organisations of which you are a member, or for a fee you can register with an agency that produces directories listing would-be home-swappers. Two such agencies are Intervac and Homelink International.

Both organisations publish directories that give a wide range of locations: houses and flats may be situated as near as Devon and Cornwall or in far away exotic locations such as the Caribbean and Hawaii. Stays can be arranged from a few weeks to a few months and, subject to satisfactory insurance, you can agree in some cases to swap cars too. Some home-swappers arrange to look after each other's pets. As most of the properties will be in residential areas, contact with neighbours and other residents will give you an excellent opportunity to learn about people living in other cultures. All you have to do is register an entry that describes your residence; the agency will give full guidelines about making contact with prospective home-swappers. You will also have to provide interested parties with information about your area: include such things as a tourist map, details of historic buildings and museums, local sports facilities, restaurants and pubs – and anything else you may think a foreign visitor might like to know.

Your home must be clean and tidy. It's a good idea if you leave simple instructions on how to use appliances and about public transport as well as a set of emergency phone numbers. Sort out who will pay the gas, electricity and phone bills during the exchange. It's also a good idea to lock away all items of value or any breakables.

Obviously, exchanging your home is an act of trust but many house-swappers are so happy with the arrangement that they return year after year. It can make your holiday money go further and enable you to visit places normally out of your price range.

Life on the ocean waves

There is a wide range of package holiday cruises to choose from, whether you're single or in a group – from study tours of archaeological sites in the Mediterranean or island hopping in the Caribbean to the full works of a round the world trip. The main organisations offering cruises are Cunard, P & O and Page and Moy. Prices are high – the most economical way to travel is in a four-berth cabin – but on the whole quality is assured, with on-board entertainment, sports and medical facilities. Requirements such as special diets can usually be catered for.

Another possibility is not luxurious – the standards are better described as clean and comfortable – but you're guaranteed some spectacular views if you take one of the Norwegian Coastal Voyage Company's mail and cargo ships from Bergen and sail 2,500 miles up Norway's coast past the Arctic Circle. Senior citizens are offered reductions in autumn, winter and spring, and the scenic journey will take you to many places that are unreachable by land. The company has recently added one or two new ships that are almost cruise standard, but prices are accordingly higher.

If, however, you would prefer not to embark on a package holiday cruise, consider a passenger-carrying cargo vessel. Only a few passengers can be carried so you won't find a wide range of facilities, but the standards of food and accommodation are usually high. It can also be cheaper than a luxury cruise, though don't expect to get to, say, Australia or the West Indies for only a few pounds! Sailings can be booked to most places in the world, such as ports in the Americas and Africa (there's even a slow boat to China) but you must be prepared to be flexible. Sailing schedules and ports of call can be affected if the weather is bad or if the cargo has been delayed. There are no doctors on board and passengers over 65 must have a certificate from their GP to say they are able to travel.

8.
The world at a glance

When you are travelling abroad, being older is not necessarily a disadvantage. Outside the youth-obsessed countries of the Western world, older people are more often than not regarded with both reverence and respect. The world really can be your oyster so long as you travel sensibly and, above all, respect and try to blend in with the cultures you visit.

Included in this chapter is a selection of countries – some popular destinations, others more unusual places that are worth considering – that can be visited exclusively or as part of a longer journey. They are divided by continent with the exception of the Russian Federation which is included under Europe because, although most of its land mass is in Asia, it has always been treated historically as a European country. Hawaii is included under the continent it is located in, Australasia; although it is one of the United States of America, it is in every other sense a Pacific island.

Each entry has a brief summary of what that country has to offer, when is the best time to go, and other useful information. Accessibility for disabled travellers varies throughout the world, so always check with the relevant tourist office or your travel agent before making a booking. And make sure that you have made your needs clear to them. It's impossible to be exact about prices, but a very rough guide to the type of cost has been given to facilitate choice: £ = budget, ££ = moderate and £££ = expensive.

If you have any doubts about your personal safety in the light of natural disaster, terrorist activity, epidemic etc, contact the Foreign Office helpline (see p 125) for up-to-date information. BBC Teletext also carries Foreign Office travel advice, as well as other holiday information.

All telephone numbers beginning with the prefix 0891 are charged at Premium rates.

Europe

Austria

CAPITAL Vienna

CURRENCY Austrian schilling

OFFICIAL LANGUAGE German. English is quite widely spoken, especially in the Tyrol

CLIMATE Warm in summer, cold in winter

BEST TIME TO VISIT Spring and summer

WHAT TO SEE Austria's magnificent baroque capital of Vienna is the European centre of classical music and opera, and no city in the world offers more exhibitions, concerts and events. Music lovers will also enjoy visiting the elegant cathedral city of Salzburg, once the home of Mozart, and the Austrian Tyrol with its towering Alpine peaks and valleys is good for both walking and motoring holidays.

AVERAGE LENGTH OF FLIGHT 2 hours

COST £

NATIONAL TOURIST OFFICE
Austrian National Tourist Office
30 St George Street
London W1R 0AL
Tel: 0171-629 0461

Belgium

CAPITAL Brussels

CURRENCY Belgian franc

OFFICIAL LANGUAGE Flemish and French. English is widely spoken in the cities

CLIMATE Warm in summer, cool and damp in winter

BEST TIME TO VISIT April to June and September to October

WHAT TO SEE Good places to stay include beautiful and historic cities such as Antwerp, which has many theatres and museums. The mediaeval walled city of Bruges and the fortress town of Ghent are also worth visiting, while nature lovers might prefer touring the forests and hills of the Ardennes.

AVERAGE LENGTH OF FLIGHT
55 minutes (or 3 hours 15 minutes from London by Eurostar)

COST £

NATIONAL TOURIST OFFICE
Belgian Tourist Office
29 Princes Street
London W1R 7RG
Tel: 0891 887799
(24-hour automated service)

Channel Islands

Guernsey

CAPITAL St Peter Port

CURRENCY Pound sterling

OFFICIAL LANGUAGE English

CLIMATE Warm in summer, windy and wet in winter

BEST TIME TO VISIT March to October; July and August can be crowded

WHAT TO SEE Guernsey offers beautiful walks and scenery and easy access to beaches. There are many rare birds and plants as well as ancient monuments and German fortifications left over from World War II. Nearby islands such as Sark, which has no cars, are just a short ferry trip away.

AVERAGE LENGTH OF FLIGHT 1 hour (sea-crossing from Weymouth 2 hours (fast), 9½ hours (slow))

COST £

NATIONAL TOURIST OFFICE
Guernsey Tourist Board
PO Box 23
St Peter Port
Guernsey GY1 3AN
Tel: 01481 723552

Jersey

CAPITAL St Helier

CURRENCY Pound sterling

OFFICIAL LANGUAGE English

CLIMATE Warm summers and cool winters

BEST TIME TO VISIT May to September; can be crowded July and August

WHAT TO SEE The largest of the Channel Islands, Jersey's many attractions include castles and the award-winning Jersey Museum which provides a journey through the island's history from ancient times to the German occupation in World War II. Renowned for its wild flowers, there is a floral festival in July and a Battle of the Flowers in August.

AVERAGE LENGTH OF FLIGHT 40 minutes (sea-crossing from Weymouth 3½ hours (fast) 11 hours (slow))

NATIONAL TOURIST OFFICE
Jersey Tourism
38 Dover Street
London W1X 3RB
Tel: 0171-493 5278

Cyprus, Republic of

CAPITAL Nicosia

CURRENCY Cyprus pound

OFFICIAL LANGUAGE Greek. English is widely spoken

CLIMATE Summers are hot, winters warm

BEST TIME TO VISIT Any time of year, but winter is the quietest period

WHAT TO SEE The walled city of Nicosia is full of museums and interesting buildings. The pretty harbour town of Paphos lies close to several impressive archaeological sites, which are easily accessible to the reasonably fit. Those wanting to get away from it all will enjoy walking in the pine forests at the foothills of the Troodos Mountains.

AVERAGE LENGTH OF FLIGHT 4 hours 30 minutes

COST £

NATIONAL TOURIST OFFICE
Cyprus Government Tourist Office
213 Regent Street
London W1R 8DA

Tel: 0171-734 9822 (Recorded information: 0891 887744)

Cyprus (northern)

CAPITAL Nicosia

CURRENCY Turkish lira

OFFICIAL LANGUAGE Turkish. Some English is spoken.

CLIMATE Hot summers, very mild winters

BEST TIME TO VISIT Any time of year; spring is becoming increasingly popular as the island is bright with blossom during March and April

WHAT TO SEE Northern Cyprus has a more peaceful atmosphere than the south, as the tourist industry is less developed and there are fewer visitors. Turkish Nicosia is an interesting town with a distinctive eastern air, and the harbour town of Kyrenia is famous for its castle. Perhaps the most celebrated landmark is Famagusta old town, a preserved Venetian town with many beautiful buildings, some of them open to visitors.

NB Northern Cyprus is technically occupied territory and is not recognised internationally or by the government of the Republic of Cyprus. Although visitors can cross the border from south to north Cyprus (provided they return the same day), the reverse is not possible.

AVERAGE LENGTH OF FLIGHT 5 hours 45 minutes (includes an hour's stopover in Turkey because there are no direct air links from the UK)

COST £

NATIONAL TOURIST OFFICE
North Cyprus Tourist Office
28 Cockspur Street
London SW1Y 5BN
Tel: 0171-930 5069

Czech Republic

CAPITAL Prague

CURRENCY Koruna

OFFICIAL LANGUAGE Czech. English is quite widely spoken in Prague

CLIMATE Hot summers and long, cold winters

BEST TIME TO VISIT Any time, but the weather is most pleasant in spring and autumn

WHAT TO SEE The magnificent capital city of Prague, with its mix of architectural styles ranging from mediaeval to nineteenth century, is unmissable. Music lovers might like to attend the 'Prague Spring', a classical music festival held in May. The Czech Republic has many fairytale castles, most of which are easily accessible by organised tours. Or take the waters at one of the many spa towns, including the most beautiful, Karlovy Vary (Carlsbad).

AVERAGE LENGTH OF FLIGHT 1 hour 45 minutes

COST £

NATIONAL TOURIST OFFICE
Czech Tourist Board
95 Great Portland Street
London W1N 5RA
Tel: 0171-436 8200

Denmark

CAPITAL Copenhagen

CURRENCY Danish krone

OFFICIAL LANGUAGE Danish. English is widely spoken

CLIMATE Cool summers and cold winters

BEST TIME TO VISIT Spring and summer

WHAT TO SEE Copenhagen has three fine palaces and the famous Tivoli gardens with its floral displays and funfair. Popular with visitors is a trip to the home of fairytale author Hans Christian Andersen in the old town of Odense on Fyn Island. The central part of the Danish mainland is known as the Garden of Denmark, and is an area of beautiful, gentle scenery that is good for motoring.

AVERAGE LENGTH OF FLIGHT 1 hour 50 minutes

COST ££

NATIONAL TOURIST OFFICE
Danish Tourist Board
55 Sloane Street
London SW1X 9SY
Tel: 0171-259 5959
(0891 600109, 24-hour)

England

CAPITAL London

CURRENCY Pound sterling

OFFICIAL LANGUAGE English

CLIMATE Warm summers and cool winters

BEST TIME TO VISIT Spring to autumn

WHAT TO SEE Everyone has their favourite spots to visit, but for a change try a city break to places such as Georgian Bath or the majestic cathedral city of York. Coach tours to the gentle Cotswold hills are also popular, as are trips to the more dramatic scenery of the Lake District. For a holiday by the sea, the Devon and south Cornwall coastline is enjoyable in spring or autumn.

COST £

NATIONAL TOURIST OFFICE
English Tourist Board
Blacks Road
London W6 9EL
Tel: 0181-846 9000

France

CAPITAL Paris

CURRENCY Franc

OFFICIAL LANGUAGE French. English is quite widely spoken in Paris, but not in the countryside

CLIMATE Warm in the summer; winters usually mild, especially in the south

BEST TIME TO VISIT Late spring and early autumn, to avoid the crowded peak season

WHAT TO SEE Paris is the cultural centre of the country, and also convenient for visiting the royal palace of Versailles and its gardens. Some of the tours to see the chateaux of the Loire Valley include wine-tastings. For those who would like to get away from it all, the colourful Provence countryside or the more rugged and dramatic coast of Brittany offer good motoring holidays or, for the very fit, walking.

AVERAGE LENGTH OF FLIGHT 1 hour 5 minutes (or 3 hours by Eurostar from London)

COST £ (can be ££ in Paris)

NATIONAL TOURIST OFFICE
French Tourist Office
179 Piccadilly
London W1V 0AL
Tel: 0891 244123

Germany

CAPITAL Berlin

CURRENCY Deutsche Mark

OFFICIAL LANGUAGE German. English is quite widely spoken, especially in cities

CLIMATE Warmest in the south; winters tend to be cold and damp, with plenty of snow in the south

BEST TIME TO VISIT The weather and the landscape are at their best from May to October

WHAT TO SEE Cruises on the Rhine are a relaxing way to see some of Germany's most beautiful and dramatic scenery. Also worth considering are wine-tasting tours in the lush vineyards of the Moselle Valley. Ramblers might enjoy exploring the vast Black Forest in a walking tour, and Ludwig II's spectacular nineteenth-century castles in Bavaria can be visited all year round.

AVERAGE LENGTH OF FLIGHT 1 hour 40 minutes

COST £

NATIONAL TOURIST OFFICE
German National Tourist Office
Nightingale House
65 Curzon Street
London W1Y 7PE
Tel: 0891 600100

Greece & Greek Islands

CAPITAL Athens

CURRENCY Drachma

OFFICIAL LANGUAGE Greek. English is quite widely spoken

CLIMATE Hot in summer, warm in winter

BEST TIME TO VISIT Spring or autumn. Summer is very hot and Greece and the islands get very crowded

WHAT TO SEE Greece and its islands provide relaxing beach holidays and historical tours. Athens is home to the famous Acropolis, but the mainland and islands also teem with archaeological sites, and visitors will have no difficulty finding tours. Of the islands, Rhodes is well known for its lovely beaches and mediaeval Crusaders City, and Crete for its mountain scenery. Facilities are better on the larger islands, but the ferry service is good and getting to the smaller islands with their quieter beaches is relatively easy.

AVERAGE LENGTH OF FLIGHT 3 hours 15 minutes

COST £

NATIONAL TOURIST OFFICE
National Tourist Organisation of Greece
4 Conduit Street
London W1R 0DJ
Tel: 0171-734 5997

Hungary

CAPITAL Budapest

CURRENCY Forint

OFFICIAL LANGUAGE Hungarian. English is not widely spoken

CLIMATE Warm summers and cold winters

BEST TIME TO VISIT May to September

WHAT TO SEE The most popular place to stay is the capital Budapest, a city with many grand historic buildings and bridges, especially in spring when a music festival takes place. Hungary has many hot thermal springs, and spa holidays are widely available. The best and most unspoilt scenery is in the Danube Bend region, best visited in springtime.

AVERAGE LENGTH OF FLIGHT 2 hours 40 minutes

COST ££

NATIONAL TOURIST OFFICE
Hungarian National Tourist Board
PO Box 4336
London SW18 4XE
Tel: 0891 171200

Iceland

CAPITAL Reykjavik

CURRENCY Iceland krona

OFFICIAL LANGUAGE Icelandic. English is widely spoken

CLIMATE Cool summers and cold winters

BEST TIME TO VISIT Summer, when the weather is at its warmest

WHAT TO SEE Reykjavik, with its many historic museums and buildings, is the cultural centre and the most popular place to stay. In July and August there are presentations of traditional dances, songs and saga readings, usually in English. Iceland abounds in spectacular waterfalls and active geysers, which can be reached by coach tours from Reykjavik. More adventurous travellers might try a trip to the north of Iceland, where there is a wealth of unspoilt scenery.

AVERAGE LENGTH OF FLIGHT 2 hours 50 minutes

COST ££

NATIONAL TOURIST OFFICE
Iceland Tourist Information Bureau
172 Tottenham Court Road
London W1P 9LG
Tel: 0171-388 7550

Ireland

CAPITAL Dublin

CURRENCY Irish punt

OFFICIAL LANGUAGE Gaelic and English

CLIMATE Warm summers, cold and wet winters

BEST TIME TO VISIT Summer

WHAT TO SEE Ireland's top city for visitors is the beautiful capital Dublin, with its famous castle and Georgian squares. Other popular historic towns include ninth-century Waterford, renowned for its crystal glass, Limerick and Tralee. Touring Ireland by coach is an ideal way of combining many of the famous landmarks, including the lakes of Killarney and Galway Bay, and enjoying the lush beauty of the Irish countryside.

AVERAGE LENGTH OF FLIGHT 50 minutes

COST £

NATIONAL TOURIST OFFICE
Irish Tourist Board
150 New Bond Street
London W1Y 0AQ
Tel: 0171-493 3201

Italy

CAPITAL Rome

CURRENCY Italian lira

OFFICIAL LANGUAGE Italian. English is spoken in cities and resorts

CLIMATE Summers are hot, but winters can be cold

BEST TIME TO VISIT April to October

WHAT TO SEE Rome, with its magnificent basilica and ancient monuments such as the Coliseum, the romantic canal city of Venice and Florence, shrine to the Italian Renaissance, all merit individual trips as there is so much to see. One of the most naturally beautiful areas of Italy is the Lakes region of Lombardy, popular with walkers; boat tours are also available. In the south, the Sorrento Peninsula with its cliff-top resorts offers glorious coastal scenery, and boat trips can be taken to the islands of Capri and Ischia.

AVERAGE LENGTH OF FLIGHT 2 hours 30 minutes

COST £ or ££

NATIONAL TOURIST OFFICE
Italian State Tourist Office
1 Princes Street
London W1R 8AY
Tel: 0171-408 1254

Luxembourg

CAPITAL Luxembourgville

CURRENCY Luxembourg franc; Belgian francs are also legal tender

OFFICIAL LANGUAGE German and French. English is quite widely spoken

CLIMATE Warm summers and cool winters

BEST TIME TO VISIT May to September

WHAT TO SEE Almost all visitors to tiny Luxembourg stay in the capital, Luxembourgville, a cathedral city of cobbled streets and ancient fortifications. The adventurous traveller might try motoring in the Luxembourg Ardennes, an area of hills, valleys and castles, where there is a museum of World War II memorabilia in Diekirch.

AVERAGE LENGTH OF FLIGHT 1 hour

COST £

NATIONAL TOURIST OFFICE
Luxembourg Tourist Office
122 Regent Street
London W1R 5FE
Tel: 0171-434 2800

Malta

CAPITAL Valletta

CURRENCY Maltese lira

OFFICIAL LANGUAGE Maltese. English is very widely spoken

CLIMATE Hot in summer, warm in winter

BEST TIME TO VISIT Late autumn to early spring

WHAT TO SEE Malta is a favourite winter destination for British travellers, and has quite a British atmosphere. Its most famous landmark is the castle of the Knights of St John in Valletta. There are several coastal resort areas but the best beaches are found on the north coast. Boat trips to the nearby picturesque island of Gozo are easily arranged.

AVERAGE LENGTH OF FLIGHT 3 hours

COST £

NATIONAL TOURIST OFFICE
Malta Tourist Office
36 Piccadilly
London W1V 0PP
Tel: 0171-292 4900

Netherlands

CAPITAL Amsterdam

CURRENCY Guilder

OFFICIAL LANGUAGE Dutch. English is very widely spoken

CLIMATE Fairly warm summers, mild damp winters

BEST TIME TO VISIT April to October

WHAT TO SEE The capital, Amsterdam, with its famous canals and Dutch gabled houses, is a good place for a short break. Its plentiful museums and galleries, including the famous Rijksmuseum and Van Gogh gallery, make it a favourite with art lovers. Boat trips on the canals are a good way to get to know the city. The Netherlands is also famous for its flowers, and coach trips are readily available to see the bulb fields in spring.

AVERAGE LENGTH OF FLIGHT 1 hour 5 minutes

COST £

NATIONAL TOURIST OFFICE
Netherlands Board of Tourism
25 Buckingham Gate
PO Box 523
London SW1E 6NT
Tel: 0891 200277

Norway

CAPITAL Oslo

CURRENCY Norwegian krone

OFFICIAL LANGUAGE Norwegian. English is widely spoken

CLIMATE Warm, short summers and cold winters

BEST TIME TO VISIT Late spring and summer

WHAT TO SEE By far the biggest attraction is Norway's western fjordlands. A motoring tour is a good way to see them, and some companies offer cruises on the fjords. Popular cities to stay in include mediaeval Bergen and Trondheim. Boat trips to the North Cape can be booked all year round, but in the summer travellers can experience the midnight sun – when the sun dips only briefly below the horizon.

AVERAGE LENGTH OF FLIGHT 1 hour 45 minutes

COST ££ or £££

NATIONAL TOURIST OFFICE
Norwegian Tourist Board
Charles House
5–11 Lower Regent Street
London SW1Y 4LR
Tel: 0171-839 6255

Northern Ireland

CAPITAL Belfast

CURRENCY Pound sterling

OFFICIAL LANGUAGE English

CLIMATE Warm summers and warm winters

BEST TIME TO VISIT May and September

WHAT TO SEE Because the country is only about the size of Yorkshire, all the main attractions are easily accessible. The wide range of scenery includes the spectacular northern coastline with the Giant's Causeway and gentler Glens of Antrim in the north east, as well as the UK's largest lake, Lough Neagh, castles and stately homes and beautiful old towns such as Derry and Armagh. Special interest holidays available include birdwatching, painting and crafts.

AVERAGE LENGTH OF FLIGHT 1 hour

COST £

NATIONAL TOURIST OFFICE
Northern Ireland Tourist Board
11 Berkeley Street
London W1X 5AD
No personal callers.
Tel: 0171-355 5040
or Freephone 0800 282662

Poland

CAPITAL Warsaw

CURRENCY Zloty

OFFICIAL LANGUAGE Polish. English is not widely spoken

CLIMATE Variable, but usually warm in summer and very cold in winter

BEST TIME TO VISIT Late spring or early autumn. Most Poles take their holidays in July and August, and resorts get very crowded

WHAT TO SEE Krakow is Poland's most beautiful old city, and home to the famous castle of Wawel. Warsaw has been carefully restored since World War II and makes an interesting visit. Ramblers will find walking in the forested foothills of the Carpathian Mountains the best way of seeing the most unspoilt areas of Poland. Another region of immense natural beauty is the Great Mazurian Lakes area, where boat trips can be taken from the lakeside resorts.

AVERAGE LENGTH OF FLIGHT 2 hours 30 minutes

COST £

NATIONAL TOURIST OFFICE
Polish National Tourist Office
310–312 Regent Street
London W1R 5AJ
Tel: 0171-580 8811

Portugal

CAPITAL Lisbon

CURRENCY Escudo

OFFICIAL LANGUAGE Portuguese. English is not widely spoken except in cities and coastal resorts

CLIMATE Hot summers and mild winters

BEST TIME TO VISIT Late spring or early autumn; coastal resorts on the Algarve can get very crowded in summer

WHAT TO SEE There is plenty to see in hilly Lisbon, including the old Moorish part of the city, a mediaeval castle and a sixteenth-century monastery. The town of Fatima is Portugal's world-famous place of pilgrimage. For beach holidays, the Algarve in the south of the country offers the best sands and the warmest weather, and a chance to explore the nearby villages with their distinctive coloured tiled houses. In the north, the city of Oporto is a favourite with wine lovers.

AVERAGE LENGTH OF FLIGHT 2 hours 30 minutes

COST £

NATIONAL TOURIST OFFICE
Portuguese National Tourist Office
22–25A Sackville Street
London W1X 1DE
Tel: 0171-494 1441

Romania

CAPITAL Bucharest

CURRENCY Leu

OFFICIAL LANGUAGE Romanian. English is not widely spoken

CLIMATE Summers are warm, winters long and bitterly cold

BEST TIME TO VISIT May to October. Avoid the winter

WHAT TO SEE Dracula tours of Transylvania visit towns and castles associated with the fifteenth-century historical figure around whom the stories are based. Birdwatchers can stay in one of the Danube delta resorts in autumn, when hundreds of species of birds from all over the world pass through on their migration south. Another site worth visiting is Bucovina, which has fifteenth-century painted monasteries.

AVERAGE LENGTH OF FLIGHT 3 hours

COST ££

NATIONAL TOURIST OFFICE
Romanian National Tourist Office
83A Marylebone High Street
London W1M 3DE
Tel: 0171-224 3692

Russian Federation

CAPITAL Moscow

CURRENCY Rouble

OFFICIAL LANGUAGE Russian. English is not widely spoken

CLIMATE Hot summers, very cold winters

BEST TIME TO VISIT Summer

WHAT TO SEE Most visitors choose to stay in either Moscow or St Petersburg. Moscow has some of the most well known landmarks in Russia: Red Square, the Kremlin and the Cathedral of St Basil's. St Petersburg in the north is a beautiful city of canals, squares and gold-domed buildings, famous as having once been the capital of the czars. Its enormous museum of art, The Hermitage, was once their Winter Palace. Visitors should note that the streets in these two cities are very long and wide; the main street in St Petersburg is several miles long and can be exhausting to walk.

AVERAGE LENGTH OF FLIGHT 3 hours 45 minutes

COST ££

NATIONAL TOURIST OFFICE
Intourist
219 Marsh Wall
Isle of Dogs
London E14 9PD
Tel: 0171-538 8600

Scotland

CAPITAL Edinburgh

CURRENCY Pound sterling

OFFICIAL LANGUAGE English

CLIMATE Warm or cool in summer, cold in winter

BEST TIME TO VISIT Summer, early autumn or New Year

WHAT TO SEE The historic capital of Edinburgh is a favourite with many visitors; its castle holds the Scottish crown jewels. For peace and tranquillity, try a tour of the lochs in the Highlands, or venture further north to the Isle of Skye. Renowned for having some of the finest golf courses in the world, Scotland is also a traditional choice for a New Year holiday.

AVERAGE LENGTH OF FLIGHT 1 hour

COST £

NATIONAL TOURIST OFFICE
Scottish Tourist Board
19 Cockspur Street
London SW1Y 5BL
Tel: 0171-930 8661

Spain

CAPITAL Madrid

CURRENCY Peseta

OFFICIAL LANGUAGE Spanish. English is spoken only in cities and resorts

CLIMATE Varies according to region. The south is warm all year round, the central region is hot in summer but very cold in winter

BEST TIME TO VISIT Spring and autumn, when the weather in most areas is fine. Avoid crowds in August, Spain's own holiday month

WHAT TO SEE Those in search of peace and quiet will like the Galician coast and the region around Cadiz. Among the most-visited of Spain's cities are Barcelona, with its famous Gaudi architecture, and Toledo, a mediaeval cathedral city. Southern Spain is renowned for its Moorish architecture, and a favourite destination is the palace of the Alhambra in Granada. The flamenco city of Seville is renowned for its colourful Easter festival and procession.

AVERAGE LENGTH OF FLIGHT 2 hours

COST £

NATIONAL TOURIST OFFICE
Spanish Tourist Office
57–58 St James's Street
London SW1A 1LD
Tel: 0171-499 0901

Sweden

CAPITAL Stockholm

CURRENCY Swedish krona

OFFICIAL LANGUAGE Swedish. English is widely spoken

CLIMATE Short, warm summers and cold winters

BEST TIME TO VISIT June to August

WHAT TO SEE After Stockholm, Sweden's cultural centre built among islands and waterways, the two historic ports of Göteborg and Malmö are the most-visited cities. For those who want beautiful scenery and peace and quiet, touring the central Dalarna region of forests and lakes is popular. Or venture north in summer to see the midnight sun, when the land is bathed in perpetual daylight.

AVERAGE LENGTH OF FLIGHT 2 hours 30 minutes

COST ££ or £££

NATIONAL TOURIST OFFICE
Swedish Travel and
Tourism Council
11 Montague Place
London W1H 2AL
Tel: 01476 578811

Switzerland

CAPITAL Berne

CURRENCY Swiss franc

OFFICIAL LANGUAGE German, French and Italian. English is widely spoken

CLIMATE Warm in summer, cold in winter

BEST TIME TO VISIT Late spring to early autumn

WHAT TO SEE Switzerland is famous for its Alpine mountains and clear air as well as its long lakes. Enjoy a relaxing stay at the lakeside resorts of Lucerne or Geneva, or try a coach trip through the Alps, probably the best way of appreciating the picturesque Swiss landscape. The historic capital, Berne, has many spectacular seventeenth-century buildings.

AVERAGE LENGTH OF FLIGHT 1 hour 30 minutes

COST ££

NATIONAL TOURIST OFFICE
Swiss Tourist Office
Swiss Court
New Coventry Street
London W1V 8EE
Tel: 0171-734 1921

Wales

CAPITAL Cardiff

CURRENCY Pound sterling

OFFICIAL LANGUAGE Welsh and English

CLIMATE Warm in summer, cold in winter

BEST TIME TO VISIT Late spring to early autumn

WHAT TO SEE Most visitors to Wales come to see the countryside, including Snowdonia National Park in the north and the Brecon Beacons in the south, both of which are areas of great natural beauty. Caernarfon, with its famous castle, is one of the most popular towns.

COST £

NATIONAL TOURIST OFFICE
Wales Tourist Board
12 Lower Regent Street
London SW1Y 4PQ
Tel: 0171-409 0969

Asia

China

CAPITAL Beijing

CURRENCY Yuan

OFFICIAL LANGUAGE Mandarin Chinese. Not much English spoken

CLIMATE Because of the vast size of China, the climate varies greatly depending on region and altitude, but tends to be hot in summer and very cold in winter

BEST TIME TO VISIT Any time of year

WHAT TO SEE Many organised tours enable visitors to spend a few days in different areas. Beijing's best-known attraction is the Forbidden City, an intricate complex of Imperial palaces, and most travellers make an excursion to see the Great Wall of China. Cruises on the Yangtze River are an excellent way to see some of China's most imposing scenery, including the famous Three Gorges steeped in Chinese folklore and legend, and usually include interesting trips ashore.

AVERAGE LENGTH OF FLIGHT
14 hours

COST £££

NATIONAL TOURIST OFFICE
China National Tourist Office
4 Glentworth Street
London NW1 5PG
Tel: 0171-935 9427

Hong Kong

CAPITAL Hong Kong

CURRENCY Hong Kong dollar

OFFICIAL LANGUAGE Chinese and English. English is widely spoken

CLIMATE Hot and humid in summer, cool and dry in winter

BEST TIME TO VISIT October and November. Avoid the typhoon season in July and August

WHAT TO SEE The main island of Hong Kong is a colourful, bustling place of shops, restaurants, temples and markets. Western influence has left its mark in the Happy Valley region, where from October to May horse racing meetings are held. But Hong Kong also has its quieter side: take a trip across the harbour to the New Territories on the mainland and find peaceful bays, paddyfields and bird sanctuaries.

AVERAGE LENGTH OF FLIGHT 13 hours 30 minutes

COST £££

NATIONAL TOURIST OFFICE
Hong Kong Tourist Association
125 Pall Mall
London SW1Y 5EA
Tel: 0891 661188

India

CAPITAL New Delhi

CURRENCY Rupee

OFFICIAL LANGUAGE A number of Indian languages and English. Quite a large proportion of the population knows a little English

CLIMATE Warm to hot in the north, very hot and humid in the south

BEST TIME TO VISIT November to April

WHAT TO SEE Many travellers to India opt for an organised tour rather than a stay in one place. Destinations to consider include colourful and romantic Rajasthan, with its magnificent palaces and fortresses (some of which are now hotels), and the historic Hindustan region of northern India including Agra with its famous fort and the Taj Mahal. South India is famous for Hindu temples and exotic wildlife. Goa, on the west coast, is a good place for a beach holiday.

AVERAGE LENGTH OF FLIGHT 9 hours

COST ££ or £££

NATIONAL TOURIST OFFICE
India Government Tourist Office
7 Cork Street
London W1X 1PB
Tel: 0171-437 3677

Indonesia

CAPITAL Jakarta

CURRENCY Rupiah

OFFICIAL LANGUAGE Bahasa Indonesian. English is not widely spoken

CLIMATE Hot and humid; the wet (monsoon) season is from October to April, the dry season from May to September

BEST TIME TO VISIT June to September

WHAT TO SEE The best-known of Indonesia's islands is exotic Bali, known as 'the Island of the Gods'. A good choice for a beach holiday, Bali also has beautiful Hindu temples and ceremonies. Lombok, a less visited island, offers more peace and quiet. Travellers to the island of Java who are interested in history might like to visit Borobodur, the world's largest Buddhist temple, which was built in the eighth century.

AVERAGE LENGTH OF FLIGHT 20 hours 20 minutes

COST £££

NATIONAL TOURIST OFFICE
Indonesian Tourist Promotions
3 Hanover Street
London W1R 9HH
Tel: 0171-493 0030

Israel

CAPITAL Jerusalem

CURRENCY New Israel shekel

OFFICIAL LANGUAGE Hebrew and Arabic. English is quite widely spoken

CLIMATE Hot April to October; November to April warm but can be chilly in the north

BEST TIME TO VISIT Any time of year

WHAT TO SEE Holidays by the Red Sea are popular and people interested in archaeology will find plenty of interest, including Roman and Crusader ruins. Or tour Israel's biblical sites, which include Bethlehem, the Sea of Galilee and the Mount of Beatitudes. Jerusalem, a city sacred to Christians, Jews and Muslims, has many fine churches and mosques and the famous Wailing Wall. Adventurous travellers might try a short stay at a kibbutz guest house.

AVERAGE LENGTH OF FLIGHT 5 hours

COST ££

NATIONAL TOURIST OFFICE
Israel Government Tourist Office
18 Great Marlborough Street
London W1V 1AF
Tel: 0171-434 3651

Japan

CAPITAL Tokyo

CURRENCY Japanese yen

OFFICIAL LANGUAGE Japanese. English is not widely spoken except in cities

CLIMATE Hot in summer, especially in the cities; winters are cold

BEST TIME TO VISIT Spring and autumn. July and August are Japan's holiday period and crowds can be a problem

WHAT TO SEE Tokyo is a good base, and visitors can take a ride on the famous bullet train to reach the various gardens, palaces and temples in the surrounding areas. For culture enthusiasts Kyoto, Japan's old capital, is the most traditional city and home of the tea ceremony. The mediaeval town of Kamakura has a gigantic statue of Buddha, and the National Park of Hakone the finest views of mighty Mount Fuji. Japan is rich in samurai houses and castles.

AVERAGE LENGTH OF FLIGHT 11 hours 30 minutes

COST £££

NATIONAL TOURIST OFFICE
Japan National Tourist Office
Heathcote House
20 Savile Row
London W1X 1AE
Tel: 0171-734 9638

Singapore

CAPITAL Singapore

CURRENCY Singapore dollar

OFFICIAL LANGUAGE Mandarin Chinese, Malay, Tamil and English. English is widely spoken

CLIMATE Hot all year round

BEST TIME TO VISIT March to October. Avoid the monsoon season from November to February

WHAT TO SEE There is plenty to explore in crowded, cosmopolitan Singapore City, including the ethnic enclaves of Chinatown, Arab Street and Little India as well as the colonial centre of the city. Those who want beaches can cross to nearby Sentosa Island by cable car. Wildlife enthusiasts might like to join a tour to one of Singapore's nature reserves.

AVERAGE LENGTH OF FLIGHT 13 hours

COST £££

NATIONAL TOURIST OFFICE
Singapore Tourist Promotion Board
Carrington House
126–130 Regent Street
London W1R 5FE
Tel: 0171-437 0033

Sri Lanka

CAPITAL Colombo

CURRENCY Sri Lankan rupee

OFFICIAL LANGUAGE Sinhala, Tamil and English

CLIMATE Very hot in summer, cooler in winter and colder at nights all year round in the hills

BEST TIME TO VISIT July and August. Monsoons from April to June and September to January

WHAT TO SEE Kandy boasts famous Kandy dancers and Buddhist Temple of the Tooth, and the nearby botanical gardens with lots of rare plants indigenous to the island. History lovers might like to visit the ruins of one of Sri Lanka's ancient cities such as Polonnaruwa, with its huge statues of Buddha. Sri Lanka also has many wildlife reserves sheltering protected species.

AVERAGE LENGTH OF FLIGHT 13 hours 45 minutes

COST ££

NATIONAL TOURIST OFFICE
Sri Lanka Tourist Board
22 Regent Street
London SW1Y 4QD
Tel: 0171-930 2627

Thailand

CAPITAL Bangkok

CURRENCY Baht

OFFICIAL LANGUAGE Thai. English is not widely spoken except in Bangkok and tourist resorts

CLIMATE Hot, wet season from June to October

BEST TIME TO VISIT November to February

WHAT TO SEE Exotic Bangkok has an abundance of Buddhist temples and palaces to visit, a famous floating market, and is excellent for shopping, especially for silk. Thailand is renowned for its beautiful beaches, and many tour companies offer combination holidays of a few days spent in Bangkok followed by a relaxing stay at a beach resort. Adventurous and fit travellers might consider a stay in northern Thailand, where they can take escorted trips to see the traditional hill tribe villages.

AVERAGE LENGTH OF FLIGHT 13 hours

COST £££

NATIONAL TOURIST OFFICE
Thailand Tourist Office
49 Albermarle Street
London W1X 3FE
Tel: 0171-499 7679

Turkey

CAPITAL Ankara

CURRENCY Turkish lira

OFFICIAL LANGUAGE Turkish. English is not widely spoken

CLIMATE Hot summers and mild winters

BEST TIME TO VISIT April to October but can be very hot in July and August

WHAT TO SEE Istanbul, the city that was once Constantinople, has many Byzantine relics, the famous church of Hagia Sophia, the Blue Mosque and rambling Topkapi palace. It also has a wonderful bazaar selling, among other items, Turkish carpets. Sites of archaeological interest abound throughout Turkey and many tours are available, as are beach holidays along the southern coast. Cruises along the Aegean and Mediterranean coasts are popular.

AVERAGE LENGTH OF FLIGHT 3 hours 45 minutes

COST ££

NATIONAL TOURIST OFFICE
Turkish Tourist and Information Office
170 Piccadilly
London W1V 9DD
Tel: 0171-355 4207

Africa

Egypt

CAPITAL Cairo

CURRENCY Egyptian pound

OFFICIAL LANGUAGE Arabic. English is quite widely spoken in the cities

CLIMATE Hot and dry all year round

BEST TIME TO VISIT November to March

WHAT TO SEE Luxor is a good base for visiting nearby ancient Egyptian monuments such as the Valley of the Kings with its royal tombs (including that of Tutankhamun) and the colossal temple of Karnak. A good way to see Egypt is by taking a cruise along the Nile. Travellers can relax on deck between trips ashore to shop in the bazaars or to visit Egypt's famous sites, which include the pyramids and sphinx of Giza and the Aswan Dam.

AVERAGE LENGTH OF FLIGHT 4 hours 45 minutes

COST ££

NATIONAL TOURIST OFFICE
Egyptian State Tourist Office
170 Piccadilly
London W1V 9DD
Tel: 0171-493 5282

The Gambia

CAPITAL Banjul

CURRENCY Gambian dalasi

OFFICIAL LANGUAGE English

CLIMATE Hot and dry; very humid in the rainy season (June to September)

BEST TIME TO VISIT Mid-November to mid-May

WHAT TO SEE The Gambian coast has beautiful white beaches lined with palms. Most hotels offer full board and many have an excellent range of sporting facilities, from tennis to archery. Land Rover trips are available to see wildlife in the bush, and boat trips along the Gambia river visit the historical remains of old slave trading posts.

AVERAGE LENGTH OF FLIGHT 5 hours 30 minutes

COST £ or ££

NATIONAL TOURIST OFFICE
Gambian National Tourist Office
57 Kensington Court
London W8 5DG
Tel: 0171-376 0093

Kenya

CAPITAL Nairobi

CURRENCY Kenyan shilling

OFFICIAL LANGUAGE English and Swahili

CLIMATE Very hot

BEST TIME TO VISIT June to September

WHAT TO SEE Kenya is famed for its wildlife and its national parks, including the famous Masai Mara. Many travellers choose to go on safari, staying in a lodge in one of the parks, from where they can join daily game drives escorted by guides. Mombasa, with its old harbour full of Arab dhows, is the main beach resort; nearby Fort Jesus, built by the Portuguese in the sixteenth century, is easily accessible to visitors.

AVERAGE LENGTH OF FLIGHT 8 hours

COST ££ or £££

NATIONAL TOURIST OFFICE
Kenya Tourist Office
25 Brook's Mews
London W1Y 1LF
Tel: 0171-355 3144

Morocco

CAPITAL Rabat

CURRENCY Moroccan dirham

OFFICIAL LANGUAGE Arabic. English is not widely spoken

CLIMATE Hot and dry

BEST TIME TO VISI April to October

WHAT TO SEE The biggest attractions are the four Imperial cities: of these, favourite destinations include Fez with its markets and winding, narrow alleyways, and Marrakesh with its famous square and souks. Morocco also has many lovely beaches, some more crowded than others. Land Rover trips down through the desert to the edge of the Sahara are most suited to the very fit.

AVERAGE LENGTH OF FLIGHT 3 hours

COST £ or ££

NATIONAL TOURIST OFFICE
Moroccan Tourist Office
205 Regent Street
London W1R 7DE
Tel: 0171-437 0073

Seychelles

CAPITAL Victoria (Mahe)

CURRENCY Seychelles rupee

OFFICIAL LANGUAGE Creole. English is not widely spoken

CLIMATE Hot all year round

BEST TIME TO VISIT June to September

WHAT TO SEE Primarily a destination of those in search of exotic, unspoilt beaches, the Seychelles are also popular with birdwatchers because the smaller islands provide a haven for resting seabirds. Trips by boat between the various islands are readily available.

AVERAGE LENGTH OF FLIGHT 11 hours 30 minutes

COST £££

NATIONAL TOURIST OFFICE
Seychelles Tourist Office
Eros House
111 Baker Street
London W1M 1FE
Tel: 0171-224 1670

South Africa

CAPITAL Pretoria

CURRENCY Rand

OFFICIAL LANGUAGE Afrikaans and English

CLIMATE Varies according to area, but generally hot

BEST TIME TO VISIT Any time of year

WHAT TO SEE Popular holidays include stays at lodges in the national parks, from where visitors can take short safaris to view the abundant wildlife. Cape Town has many beautiful bays, offers a chance to visit Table Mountain by cable car and is also convenient for visiting the famous South African vineyards. For touring, eastern Transvaal has South Africa's most dramatic mountain scenery.

AVERAGE LENGTH OF FLIGHT Around 13 hours

COST £££

NATIONAL TOURIST OFFICE
South African Tourism Board
5/6 Alt Grove
London SW19 4DZ
Tel: 0181-944 8080

Tunisia

CAPITAL Tunis

CURRENCY Tunisian dinar

OFFICIAL LANGUAGE Arabic. French is the second language. English is not widely spoken except in resorts

CLIMATE Hot in summer, warm in winter

BEST TIME TO VISIT Spring and autumn

WHAT TO SEE The main attraction of Tunis is the remains of the ancient port of Carthage which lie just outside the city; art and archaeology tours are available which take in Carthage and Roman sites in the north of the country. Scenery ranges from oak forests in the north to desert dunes and oases in the south. Lake Tunis is a haven for birdwatchers.

AVERAGE LENGTH OF FLIGHT 2 hours 30 minutes

COST £

NATIONAL TOURIST OFFICE
Tunisian National Tourist Office
77A Wigmore Street
London W1H 9LJ
Tel: 0171-224 5561

Australasia

Australia

CAPITAL Canberra

CURRENCY Australian dollar

OFFICIAL LANGUAGE English

CLIMATE Generally hot from November to February, warm or cool from March to October

BEST TIME TO VISIT November to February – the Australian summer

WHAT TO SEE Sydney with its famous opera house is the main cultural centre. Cruises through the Great Barrier Reef are popular, usually offering travellers a chance to stop off on various islands en route. Adelaide in the south is a good base for touring the wine growing region of the Barossa Valley, and can also be the starting point for anyone wishing to try a train trip through the scorching Australian desert to the central town of Alice Springs, from where Ayers Rock can be visited.

AVERAGE LENGTH OF FLIGHT 22 hours

COST £££

NATIONAL TOURIST OFFICE
Australian Tourist Commission
Gemini House
10/18 Putney Hill
London SW15 6AA
Tel: 0181-780 2227

Hawaii (USA)

CAPITAL (STATE) Honolulu

CURRENCY US dollar

OFFICIAL LANGUAGE English

CLIMATE Hot and sunny all year round

BEST TIME TO VISIT Summer is cheapest. Winter is popular, but has more rain

WHAT TO SEE Hawaii's main island has the spectacular Volcanoes National Park, with its lava streams and craters; some of the walks are accessible to wheelchairs. For nature lovers, the island of Kauai offers mountain scenery, beautiful beaches and good birdwatching, while more adventurous travellers might like to try whalewatching on Maui.

AVERAGE LENGTH OF FLIGHT 19 hours

COST £££

NATIONAL TOURIST OFFICE
None in the UK. Contact Visit USA Association
Tel: 0891 600530 for brochures

New Zealand

CAPITAL Wellington

CURRENCY New Zealand dollar

OFFICIAL LANGUAGE English

CLIMATE Warm in the summer, mild in the winter, and generally warmer in the far north

BEST TIME TO VISIT December to March – New Zealand's summer

WHAT TO SEE The most picturesque of the big cities is Wellington, North Island, which also has some good beaches. Rotorua is the centre of Maori culture, and is also famous for its thermal springs, geysers and boiling mud-pools. Cruises are available in the Bay of Islands, one of New Zealand's most historic areas in North Island, famous for its associations with the explorer Captain Cook. New Zealand has many national parks.

AVERAGE LENGTH OF FLIGHT 30 hours

COST £££

NATIONAL TOURIST OFFICE
New Zealand Tourism Board
New Zealand House
80 Haymarket
London SW1Y 4TQ
Tel: 0839 300900

North America

Barbados

CAPITAL Bridgetown

CURRENCY Barbados dollar

OFFICIAL LANGUAGE English

CLIMATE Hot summers, warm winters

BEST TIME TO VISIT Mid-December to mid-May

WHAT TO SEE Beach lovers will find that the west coast of Barbados has some of the finest sands in the Caribbean. Further towards the island's centre, grand old plantation houses and their gardens attract many visitors. It's possible to hire a private car to get to the more remote north east of the island, where the best walking country, with flower forests and a wildlife reserve, is to be found.

AVERAGE LENGTH OF FLIGHT 7 hours 30 minutes

COST ££

NATIONAL TOURIST OFFICE
Barbados Board of Tourism
263 Tottenham Court Road
London W1P 0LA
Tel: 0171-636 9448

Canada

CAPITAL Ottawa

CURRENCY Canadian dollar

OFFICIAL LANGUAGE English and French

CLIMATE Summers usually quite warm, winters extremely cold

BEST TIME TO VISIT Late spring to early autumn

WHAT TO SEE For nature lovers, Canada has some of the world's most breathtaking wild scenery. The Great Lakes region in the east offers travellers a choice of resorts, and includes the spectacular Niagara Falls. In the west, train trips through the majestic Canadian Rocky Mountains are the best way to see both the scenery and the wildlife. For a city-based holiday, Quebec offers old French style architecture and cobbled streets.

AVERAGE LENGTH OF FLIGHT 7 hours

COST £££

NATIONAL TOURIST OFFICE
Canadian Tourist Office
Macdonald House
1 Grosvenor Square
London W1X 0AB
Tel: 0891 715000

Cuba

CAPITAL Havana

CURRENCY Cuban peso, but the US dollar is acceptable. American Express traveller's cheques cannot be used

OFFICIAL LANGUAGE Spanish. Not much English spoken

CLIMATE Hot

BEST TIME TO VISIT November to April

WHAT TO SEE Old Havana is an interesting mixture of alleyways, cobbled streets and Spanish colonial architecture; sixteenth-century Trinidad in the south is an old town that has been designated a 'living museum'. Cuba's second city, Santiago, has 14 museums, including one devoted to Bacardi rum. There are many good beaches.

AVERAGE LENGTH OF FLIGHT 9 hours

COST £ or ££

NATIONAL TOURIST OFFICE
Cuban Tourist Office
167 High Holborn
London WC1V 6PA
Tel: 0171-379 1706

Mexico

CAPITAL Mexico City

CURRENCY Mexican peso

OFFICIAL LANGUAGE Spanish. English is not widely spoken except in Mexico City

CLIMATE Hot and wet in summer, warm in winter

BEST TIME TO VISIT October to May

WHAT TO SEE Mexico is rich in archaeological sites from the Aztec and Mayan civilisations, including the pyramids at Teotihuacan and the ruins of Palenque. There are also many fine old Spanish colonial churches and other buildings, and plenty of colourful markets.

AVERAGE LENGTH OF FLIGHT 12 hours and 20 minutes

COST ££

NATIONAL TOURIST OFFICE
Mexican Tourist Office
60–61 Trafalgar Square
London WC2N 5DS
Tel: 0171-734 1058

United States of America

CAPITAL Washington DC

CURRENCY US dollar

OFFICIAL LANGUAGE English

CLIMATE Varies greatly depending on the region; generally cool in the north, warm or hot in the south

BEST TIME TO VISIT Also depends on region – but the major cities can be visited at any time of year

WHAT TO SEE For those in search of the bright lights, New York, San Francisco and Las Vegas offer a variety of cultural and big-city entertainments. Favourite trips to the deep South include visiting old plantation houses, and paddle steamer journeys on the Mississippi. Some of the most dramatic landscape is in the west, including the mile-deep Grand Canyon and Monument Valley, while the red and gold scenery of New England in the autumn provides gentler scenery. Wine enthusiasts might try a tour of the Californian Napa Valley vineyards.

AVERAGE LENGTH OF FLIGHT 5–8 hours, depending on your destination city

COST ££ or £££

NATIONAL TOURIST OFFICE
None in the UK. Contact Visit USA Association
Tel: 0891 600530 for brochures

South America

Brazil

CAPITAL Brasilia

CURRENCY Cruzeiro

OFFICIAL LANGUAGE Portuguese. English is not widely spoken except in Rio

CLIMATE Hot in most areas, and very rainy in the north

BEST TIME TO VISIT February for the carnival, otherwise any time of year

WHAT TO SEE The parades, floats and colourful costumes of the Rio carnival in February are the big attraction for many visitors, but Brazil is also a paradise for nature lovers. The Pantanal region in western Brazil, rich in tropical species, is popular for wildlife tours, some with a focus on birdwatching. Boat trips up the Amazon offer the chance to venture off the beaten track and into the rainforest, while history lovers might like to tour the colonial goldrush towns of the Minas Gerais area.

AVERAGE LENGTH OF FLIGHT 11 hours

COST £££

NATIONAL TOURIST OFFICE
Brazilian Tourist Board
32 Green Street
London W1Y 4AT
Tel: 0171-499 0877

Chile

CAPITAL Santiago

CURRENCY Chilean peso

OFFICIAL LANGUAGE Spanish. English is not widely spoken

CLIMATE Hot and sunny November to March, warm and wet April to September

BEST TIME TO VISIT November to March

WHAT TO SEE As the longest country in the world, mountainous Chile boasts a great variety of magnificent scenery, ranging from the red sand of the Atacama desert in the north to the spectacular glaciers in the south. Organised tours offer the chance to see rare species in national parks, or venture off the beaten track to camp by mountain lakes.

AVERAGE LENGTH OF FLIGHT 18 hours

COST £££

NATIONAL TOURIST OFFICE
None. Contact
Chilean Embassy
12 Devonshire Street
London W1N 2DS
Tel: 0171-580 6392

Peru

CAPITAL Lima

CURRENCY Nuevo sol

OFFICIAL LANGUAGE Spanish and Quechua. English is not widely spoken

CLIMATE On the coast, hot and sunny December to March and warm April to November; in the mountains, hot and rainy December to March but warm and dry June to November

BEST TIME TO VISIT December to April

WHAT TO SEE Peru is a mountainous country, and many visitors prefer to take an organised tour. The biggest attraction for travellers is the mysterious remains of the Inca city, Machu Picchu, which is perched atop a mountain and accessible by train. The adventurous might like to try a flight over the famous Nazca Lines. Lake Titicaca is a popular stop, where it is possible to take boat trips to see the villages on their reed islands.

AVERAGE LENGTH OF FLIGHT 14 hours

COST £££

NATIONAL TOURIST OFFICE
Peruvian Embassy
Tourist Department
52 Sloane Street
London SW1X 9SP
Tel: 0171-235 2545

USEFUL ADDRESSES

Listed below are the addresses and, where available, telephone numbers of organisations mentioned in the book. Telephone numbers beginning with 0891 are charged at Premium rates.

Assistance in arranging trips

Marco Polo Travel Advisory Service
24A Park Street
Bristol BS1 5JA
Tel: 0117-929 4123

Trailfinders
42–48 Earl's Court Road
London W8 6FT
Tel: 0171-938 3366

Wexas International
45–49 Brompton Road
London SW3 1DE
Tel: 0171-589 3315

Associations

Royal Geographical Society
1 Kensington Gore
London SW7 2AR
Tel: 0171-589 5466
Holds regular lectures and discussions on a wide variety of geographical subjects

British Tourist Authority
Thames Tower
Blacks' Road
London W6 9EL
Tel: 0181-846 9000

Camping and caravaning

Camping and Caravaning Club
Green Fields House
Westwood Way
Coventry CV4 8JH
Tel: 01203 694995

The Caravan Club
East Grinstead House
East Grinstead
West Sussex RH19 1UA
Tel: 0134-232 6944

Eurocamp
28 Princess Street
Cheshire WA16 6BU
Tel: 01565 626262

National Caravan Council
Caravan Holiday Home
Information Bureau
Department 203
PO Box 26
Lowestoft NR32 3OH

SUPPLIERS OF EQUIPMENT
YHA Adventure Shop
14 Southampton Street
London WC2E 7HY
Tel: 0171-836 8541

Companions

Especially for Women
c/o **Single Again**
Suite 33
10 Barley Mow Passage
London W4 4PH
Tel: 0181-749 3745

Travel Companions
110 High Mount
Station Road
London NW4 3ST
Tel: 0181-202 8478

Women-Welcome-Women
88 Easton Street
High Wycombe
Bucks HP11 1LT
Tel: 01494 465441

Courier flights

British Airways Travel Shops
World Cargo Centre
Express Cargo Terminal S126
Heathrow Airport
Hounslow TW6 2JS
Tel: 0181-564 7009
For details of BA's courier flights

Cruising

The Cruise Advisory Travel
35 Blue Boar Row
Salisbury
Wilts SP1 1DA
Tel: 01722 335505
Advice on cruising world-wide

Cunard
South Western House
Canute Road
Southampton SO14 3NR
Tel: 01703 716500

Norwegian Coastal Voyage Company
15 Berghem Mews
Blythe Road
London W14 0HN
Tel: 0171-371 4011

P&O
77 New Oxford Street
London WC1A 1PP
Tel: 0171-800 2222

Page & Moy
136–140 London Road
Leicester LE2 1EN
Tel: 0116-250 7000

Wieder Travel
Charing Cross Shopping
Concourse
Strand
London WC2N 4HZ
Tel: 0171-836 6363

Home swap

Homelink International
Linfield House
Gorse Hill Road
Virginia Water
Surrey GU25 4AS
Tel: 01344 842642

Intervac
6 Siddlas Lane
Allestree
Derby DE3 2DY
Tel: 01225 892208

Hostels

Youth Hostel Association
YHA Groups Adviser
Westbury High Street
Napton
Rugby
Warwickshire CV23 8LZ
Tel: 01926 815169

YMCA Centre
Lakeside
Ulverston
Cumbria LA12 8BD
Tel: 01539 531758

Insurance

The Association of British Insurers
51 Gresham Street
London EC2V 7HQ
Tel: 0171-600 3333

Maps and guides

AA Publications
Automobile Association
Fanum House
Basingstoke
Hants RG21 4EA
Tel: 01256 491513
Guides and maps

Lonely Planet Publications
Devonshire House
12 Barley Mow Passage
London W4 4PH
Tel: 0181-742 3161
One of the world's largest guide book publishers

Michelin Guides
Edward Hyde Building
38 Clarendon Road
Watford WD1 1SX
Tel: 01923 415000
Publishes the famous red and green guide books

The National Map Centre
22–24 Caxton Street
London SW1H 0QU
Tel: 0171-222 2466
Main agent for Ordnance Survey maps

RAC Publishing
RAC House
PO Box 100
South Croydon
Surrey CR2 6XW
Tel: 0181-686 0088
Guides, handbooks and maps for UK and continent

Rough Guides
1 Mercer Street
London WC2H 9QJ
Tel: 0171-379 3329
Publishes the Rough Guide series of travel books

Stanfords
12–14 Long Acre
London WC2E 9LP
Tel: 0171-836 1321
Stocks a full range of guide books and is the largest map seller in the world. Offers a mail order and telephone service

Trailfinders
Travel Centre
42–48 Earl's Court Road
London W8 6FT
Tel: 0171-938 3366
Guides, maps and travel equipment

Vacation Work Publications
9 Park End Street
Oxford OX1 1HJ
Tel: 01865 241978
Books for budget travellers and anyone wanting to work abroad

YHA Bookshops
14 Southampton Street
London WC2E 7HY
Tel: 0171-836 8541
Books, maps and guides for hostellers and budget travellers

Medical

British Airways Travel Clinic
156 Regent Street
London W1R 6LB
Tel: 0171-434 4700

Vaccination Clinic
Inoculation advice and treatment only)
46 Wimpole Street
London W1M 1DG
Tel: 0171-486 3665

Recorded information
Tel: 0171-831 5333

Department of Health
Richmond House
79 Whitehall
London SW1A 2NS
Tel: 0171-210 4850
(General enquiries)

Intermedic
777 Third Avenue
New York NY 10017, USA

International Association for Medical Assistance for Travellers (IAMAT)
Dept TH
417 Center Street
Lewiston NY 14092, USA
Tel: 001 (716) 754 4883

MASTA
c/o London School of Hygiene and Tropical Medicine
Keppel Street
London WC1E 7HT
Tel: 0171-631 4408 (Medical Advisory Service)
0891 224100 (Malaria Helpline)

Red Cross
Look in your local telephone directory

MEDICAL (SPECIAL NEEDS)

Air Transport Users Council
5th Floor, 103 Kingsway
London WC2B 6QX
Tel: 0171-242 3882

Department of Transport
76 Marsham Street
London SW1P 4DR
Tel: 0171-276 3000

Disabled Living Centres Council
Winchester House
11 Cranmer Road
London SW9 6EJ
Tel: 0171-820 0567

Disabled Living Foundation
380–384 Harrow Road
London W9 2HU
Tel: 0171-289 6111

Holiday Care Service
2nd Floor, Imperial Buildings
Victoria Road
Horley
Surrey RH6 7PZ
Tel: 01293 774535

Royal Association for Disability and Rehabilitation (RADAR)
12 City Forum
250 City Road
London EC1V 8AF
Tel: 0171-250 3222

Royal National Institute for the Blind (RNIB)
224 Great Portland Street
London W1N 6AA
Tel: 0171-388 1266

Tripscope
The Courtyard
Evelyn Road
London W4 5JL
Tel: 0181-994 9294

Motoring

MOTORING ORGANISATIONS

The Automobile Association
Norfolk House
Priestley Road
Basingstoke
Hants RG24 9JR
Tel: 01256 21023

The Royal Automobile Club
PO Box 100
RAC House
7 Brighton Road
South Croydon
Surrey CR2 6XW
Tel: 0181-686 0088

DISABILITY

Department of Transport
Mobility Unit
1st Floor
Great Minster House
76 Marsham Street
London SW1P 4DR
Tel: 0171-271 5257

Disabled Drivers Association
Ashwellthorpe Hall
Norwich
Norfolk NR16 1EX
Tel: 01508 489449

Disabled Drivers Motor Club
Cottingham Way
Thraxton
Northants NN14 4PL
Tel: 01832 734724

Disabled Motorists Federation
Unit 2A
Atcham Estate
Shrewsbury SY4 4UG
Tel: 01743 761181

CAR RENTALS

Avis Rent-a-Car
Avis House
Park Road
Bracknell
Berks RG12 2EW
Tel: 01344 426644

Hertz Rent-a-Car
Radnor House
1272 London Road
London SW16 4XW
Tel: 0181-679 1777

Passport

Passport Office
Clive House
70–78 Petty France
London SW1H 9HD
Tel: 0990 210410

Political unrest

Foreign Office Advice Line
The Consular Department
The Foreign and
Commonwealth Office
1 Palace Street
London SW1E 5HE
Tel: 0171-238 4503
Gives up-to-date advice on the political and social stability of countries. The unit also issues useful free leaflets

Pressure group

Single Travellers Action Group (STAG)
Church Lane
Charnbrook
Bedford MK44 1HR

Campaigns for the rights of single travellers and issues quarterly newsletter listing supplement-free holidays. Annual subscription is £5

Rail

For rail travel in the UK, contact your nearest railway enquiry office (in the telephone book)

Eurostar
Eurostar Enquiries
Waterloo Station
London SE1 8SE
Tel: 0345 881881

Renting

Blakes Country Cottages
Stony Bank Road
Spring Hill
Earby
Lancs BB8 6RN
Tel: 01282 445225

English Country Cottages
Grove Farm Barns
Fakenham
Norfolk NR21 9NB
Tel: 01328 864292

The Landmark Trust
Shottesbrooke
Maidenhead
Berks SL6 3SW
Tel: 01628 825925

The National Trust
36 Queen Anne's Gate
London SW1H 9AS
Tel: 0171-222 9251

For a copy of the National Trust's 'Holiday Cottages' brochure send a cheque or postal order for £1 to National Trust (Enterprises) Ltd, PO Box 536, Melksham, Wilts SN12 8SX (Tel: 01225 791133)

Retreats

The Buddhist Society
58 Eccleston Square
London SW1V 1PH
Tel: 0171-834 5858

Cruse – Bereavement Care
126 Sheen Road
Richmond
Surrey TW9 1UR
Tel: 0181-940 4818

National Council of Hindu Temples
559 St Albans Road
Watford
Herts WD2 6JH
Tel: 01923 6784168

National Retreat Association
Central Hall
256 Bermondsey Street
London SE1 9UJ
Tel: 0171-357 7736

Security

Crime Prevention Officer at your local police station

Animal Aunts
Wydwooch
45 Farview Road
Headley Down
Hants GU35 8HQ
Tel: 01428 712611

Homesitters
Buckland Wharf
Buckland ·
Aylesbury
Bucks HP22 5LQ
Tel: 01296 630730

Housewatch Ltd
Little London
Berden
Bishop's Stortford
Herts CM23 1BE
Tel: 01279 777412

Study/special interest holidays

Acorn Activities
PO Box 120
Hereford HR4 8YB
Tel: 01432 830083

Association of Cultural Exchange Study Tours
Babraham
Cambridge CB2 4AP
Tel: 01223 835053

Central Bureau for Educational Visits and Exchanges
10 Spring Gardens
London SW1A 2BN
Tel: 0171-389 4004

En Famille Overseas
The Old Stables
60B Maltravers Street
Arundel
West Sussex BN18 9BG
Tel: 01903 883266

Outward Bound
PO Box 1219
Windsor SL4 1XR
Tel: 01753 730060

Saga Holidays
The Saga Building
Middelburg Square
Folkestone
Kent CT20 1AZ
Tel: 01303 711111

Ski-Club of Great Britain
118 Eaton Square
London SW1W 9AF
Tel: 0171-245 1033

Solo's Holidays
54–58 High Street
Edgware
Middlesex HA8 7ED
Tel: 0181-951 2800

University of the Third Age (U3A)
1 Stockwell Green
London SW9 9JF
Tel: 0171-737 2541
Send a large sae for details of country-wide U3As

Waymark Holidays
44 Windsor Road
Slough SL1 2EJ
Tel: 01753 516477
For cross-country skiing

Travel organisations

Air Travel Organiser's Licence (ATOL)
Room T506
CAA House
45–59 Kingsway
London WC2B 6TE
Tel: 0171-832 5620/6600

Association of British Travel Agents (ABTA)
55–57 Newman Street
London W1P 4AH
Tel: 0171-637 2444
Information line 0891 202520

Association of Independent Tour Operators (AITO)
133a St Margaret's Road
Twickenham
Middlesex TW1 1RG
Tel: 0181-744 9280

Volunteer

British Trust for Conservation Volunteers
36 St Mary's Street
Wallingford
Oxon OX10 0EU
Tel: 01491 839766

Winged Fellowship Trust
Angel House
20–32 Pentonville Road
London N1 9XD
Tel: 0171-833 2594

See also the house-sitting organisations listed under 'Security'

Walking

HF Holidays
Imperial House
Edgware Road
London NW9 5AL
Tel: 0181-905 9556

Ramblers Association
1–5 Wandsworth Road
London SW8 2XX
Tel: 0171-582 6878

Ramblers Holidays
Box 43
Welwyn Garden City
Herts AL8 6PQ
Tel: 01707 331133

FURTHER READING

Bed and Breakfast Guide, published by the Automobile Association, Basingstoke (updated annually)

Flightplan, published by the Air Transport Users Committee, London (see p 124)

Flying High, published by the Disabled Living Foundation (see p 124)

Handbook for Women Travellers by M and G Moss, published by Piatkus, London, 1995

Health Advice for Travellers, booklet T5, published by the Department of Health is available free by phone (Tel: 0800 555777) or from main post offices and some travel agents

Healthy Breaks in Britain and Ireland by Catherine Beattie, published by Discovery Books in conjunction with the tourist boards of England, Scotland, Wales and Northern Ireland, 1994

Holidays in the British Isles, published by RADAR, London

Holidays and Travel Abroad, published by RADAR, London (updated annually)

Home from Home, published by the Central Bureau for Educational Visits and Exchanges

How to Start a Conversation and Make Friends by Don Gabor, published by Sheldon Press, London, 1995

Nothing Ventured: Disabled people travel the world edited by Alison Walsh, published by Harrap Columbus, London, 1991 (a Rough Guides book)

The Traveller's Handbook edited by Caroline Brandenburger, published by Wexas, London, 1994

Traveller's Health: How to stay healthy abroad by Dr Richard Dawood, published by Oxford University Press, London, 1995

Traveller's Information Special Needs, produced by Heathrow Airport, can be obtained by ringing 0181-745 7495

Visitor's Guide to Health, published by the Central Office of Information. Obtainable via the Health Literature Line: 0800 555777

ABOUT AGE CONCERN

The World at your feet: A traveller's guide for women in mid-life and beyond is one of a wide range of publications produced by Age Concern England, the National Council on Ageing. Age Concern England is actively engaged in training, information provision, fundraising and campaigning for retired people and those who work with them, and also in the provision of products and services such as insurance for older people.

A network of over 1,400 local Age Concern groups, with the support of around 250,000 volunteers, aims to improve the quality of life for older people and develop services appropriate to local needs and resources. These include advice and information, day care, visiting services, transport schemes, clubs, and specialist facilities for older people who are physically and mentally frail.

Age Concern England is a registered charity dependent on public support for the continuation and development of its work.

Age Concern England
1268 London Road
London SW16 4ER
Tel: 0181-679 8000

Age Concern Scotland
113 Rose Street
Edinburgh EH2 3DT
Tel: 0131-220 3345

Age Concern Cymru
4th Floor
1 Cathedral Road
Cardiff CF1 9SD
Tel: 01222 371566

Age Concern Northern Ireland
3 Lower Crescent
Belfast BT7 1NR
Tel: 0232 245729

PUBLICATIONS FROM AGE CONCERN BOOKS

Money Matters

Earning Money in Retirement
Kenneth Lysons
Many people, for a variety of reasons, wish to continue in some form of paid employment after they have retired. This helpful guide explores the practical implications of such a choice and highlights some of the opportunities available.
£3.99   0-86242-103-9

The Insurance Handbook: A guide for older people
Wayne Asher
Older people have particular needs – and opportunities – when purchasing insurance. This practical guide provides a useful overview of the products on the market, including home and contents, car, holiday, health and life insurance. The aim is to help readers get value for money and find a product that is really right for their needs.
£3.99   0-86242-146-2

General

Out and About: A travel and transport guide
Richard Armitage and John Taylor
A comprehensive source of information on travel and transport for older people and others with limited mobility. Whether planning a trip to the local shops or a journey abroad, this book provides a step-by-step guide to the arrangements that need to be made.
£4.66   0–86242–092–X

An Active Retirement
Nancy Tuft
Packed with information on hobbies, sports, educational opportunities and voluntary work, this practical guide is ideal for retired people seeking new ways to fill their time but uncertain where to start.
£5.33   0–86242–119–5

Looking Good, Feeling Good: Fashion and beauty in mid-life and beyond
Nancy Tuft
Positive, upbeat and awash with useful advice and ideas, this book encourages the over-50s to take pride in their appearance and challenges the popular view that interest in fashion and beauty passes with the years.
£5.33   0–86242–102–0

If you would like to order any of these titles, please write to the address below, enclosing a cheque or money order for the appropriate amount and made payable to Age Concern England. Credit card orders may be made on 0181-679 8000.

Mail Order Unit
Age Concern England
1268 London Road
London SW16 4ER

INFORMATION FACTSHEETS

Age Concern England produces over 30 factsheets on a variety of subjects. Among these the following titles may be of interest to readers of this book:

Factsheet 4   *Holidays for older people*
Factsheet 26 *Travel information for older people*
Factsheet 30 *Leisure education*

To order factsheets

Single copies are available free on receipt of a 9″ × 6″ sae. If you require a selection of factsheets or multiple copies totalling more than five, charges will be given on request.

A complete set of factsheets is available in a ring binder at a cost of £36, which includes the first year's subscription. The current cost of an annual subscription for subsequent years is £17. There are different rates of subscription for people living outside the UK.

For further information, or to order factsheets, write to:
Information and Policy Department
Age Concern England
1268 London Road
London SW16 4ER

INDEX

accidents *62*, *71*
activity holidays *23*, *83–85*
Africa *68*, *111–114*
AIDS *71*
air travel *16*, *23*, *35*, *36–37*, *42*, *71*
    courier flights *42*, *122*
    and medical conditions *66*
alone, holidaying *12–19*, *42*, *43*
animals, bites from *72–73*
asthma sufferers *66*
athlete's foot *71–72*
Australia *29*, *68*, *115*
Austria *91*

B & Bs *13–14*, *43*
backpacks *55–56*
bags, carrying *58*, *60*
bail bonds *50*
Barbados *116*
bee stings *72*
Belgium *91*
bites *72–73*
blind travellers *31*, *32*, *36*, *37*
blood transfusions *71*, *76*
booking holidays *34–35*, *41–42*
Brazil *119*

bronchitis *66*
burglaries, preventing *51–52*, *126–127*

cameras *49*, *58*; *see also* photography
camping and caravanning *85–86*, *121–122*
Canada *29*, *117*
car travel *47*, *50*, *61–62*, *71*, *125*
    and disabled travellers *29*, *38–39*
    and hiring cars *38*, *47*, *125*
care attendant agencies *33*
cash, taking *45*, *49*, *59*
cash dispensers *46*, *64*
Channel Islands *92*
Chile *119*
China *106*
cholera *68–69*
city breaks, European *22–23*
clothes *53–55*
commodes, collapsible *35*
companions, finding *19*, *20–22*
constipation *73*
contact lenses *66*
conversations, starting *19*, *26–28*
cottages, renting *82*, *126*
courier flights *42*, *122*

credit cards *44, 46, 48, 59, 78*
    loss of *63–64*
cruises *39, 87–88, 122*
Cuba *117*
cuts and grazes *73*
Cyprus *93–94*
cystitis *73*
Czech Republic *94*

Denmark *95*
dentists *67, 79*
diabetes *66*
diarrhoea *73–74*
disabilities, travellers with *29ff*
doctors *77–78*
dog bites *72–73*
drinking, safe *77*
driving licences *47*

eating, safe *77*
Egypt *111*
England *95*
escort services *33*
Eurocheques *45–46*
expenses, holiday *43–44*

families, staying with *13, 85*
feet and footwear *54, 67, 71–72*
ferries *39, 42*
first aid kits *75–76*
Foreign Office Helpline *90, 125*
France *63, 85, 96*

Gambia, The *112*
Germany *63, 96*
Greece and Greek Islands *33, 97*

green cards *50*
group tours *23–28, 31–32*;
    see also package holidays
Guernsey *92*
guest houses *43, 60, 71*
guide books *32, 41, 43, 58, 123–124*

Hawaii *89, 115*
health care *see* medical aspects
health farms *83*
heart disease *66*
heat exhaustion *74*
hepatitis A *69*
historic properties *82*
Holiday Care Service *33*
home security *51–53, 126–127*
home-swapping *86–87, 122–123*
Hong Kong *63, 107*
hostels *see* youth hostels
hotels *43, 60, 71*
house-sitting *51, 53, 81*
Hungary *47, 97*
hygiene, personal *72, 77*

Iceland *98*
immunisation *33, 67, 68–70*
India *107*
Indonesia *58, 108*
injections *34, 71*
insect bites and stings *72*
insect repellents *70*
insurance *47–49, 123*
    of house contents *51*
    medical *34, 49, 67, 78*
Ireland *98*
Ireland, Northern *101*
Israel *46, 108*
Italy *63, 99*

Japan *109*
Jersey *92*
jet lag *74*

Kenya *112*

Landmark Trust *82, 126*
languages, learning *19*
luggage *55–56, 60*
Luxembourg *99*

malaria *67, 70*
Malta *100*
maps *58, 59, 123–124, see also* guide books
medical aspects (*see also* immunisation):
   certificates *35*
   first aid kits *75–76*
   getting attention *67, 77–78*
   insurance *49, 67, 78*
   taking medication *33–34, 65–66*
   useful addresses *124–125*
meeting people *19, 26–28*
Mexico *118*
Morocco *113*
mosquitoes *68, 70, 72*
Muslim countries *18, 58*

National Trust properties *82, 126*
Netherlands *63, 100*
New Zealand *29, 116*
Northern Ireland *101*
Norway *37, 87–88, 101*

Orange Badge schemes *38*

package holidays *23–28, 43, 44, 80, 87–88*
packing *53–56*
passports *46–47, 59*
Peru *120*
pets, care of *53*
photography *17–18*
planning holidays *10–11, 29–33, 41*
Poland *102*
poliomyelitis *69*
political unrest *46, 90, 125*
Portugal *47, 102*
prescriptions, medical *33, 66*

rail travel *37–38, 62, 126*
rambling *84, 128*
relaxation *15, 16*
renting properties *82, 126*
restaurants, eating in *18, 43–44*
retreats *80–81, 126*
road travel *see* car travel
Romania *103*
Russian Federation *89, 103*

safety precautions *57ff*
Scotland *104*
Seychelles *113*
shoes *54*
Singapore *109*
single travellers *12–19, 42, 43*
skiing holidays *84–85*
skin infections *71–72*
snake bites *73*
South Africa *114*
Spain *47, 50, 63, 104*
special-interest holidays *23, 30, 83, 127*

spectacles 66
sporting holidays 83–85
Sri Lanka 110
stings, insect 72
study holidays 24, 83, 127
suitcases 56, 60
sunburn 75
Sweden 105
Switzerland 37, 105

teeth, care of 67, 79
telephones 28, 38
tetanus 70
Thailand 110
thrush 72
tiredness, combating 16
trains 37–38, 62, 126
travel organisations 127–128
travel sickness 75
traveller's cheques 45, 59

Tunisia 114
Turkey 111
typhoid fever 69

United States of America 29, 45, 118

visas 44, 47
volunteer work 81–82, 128

Wales 106
walking holidays 63, 84, 128
wasp stings 72
water, sterilising 76, 77
weekend trips 13–14
wheelchair users 29, 34, 35

yellow fever 68
youth hostels 43, 84, 123